Publishing
as a
Vocation

Publishing
as a Vocation

Studies of an Old Occupation in a New Technological Era

Irving Louis Horowitz

Routledge
Taylor & Francis Group

LONDON AND NEW YORK

First published 2011 by Transaction Publishers

2 Park Square, Milton Park, Abingdon, Oxfordshire OX14 4RN
711 Third Avenue, New York, NY 10017

Routledge is an imprint of the Taylor & Francis Group, an informa business

First issued in paperback 2017

Library of Congress Catalog Number: 2010024041

Library of Congress Cataloging-in-Publication Data

Horowitz, Irving Louis.
    Publishing as a vocation : studies of an old occupation in a new technological era / Irving Louis Horowitz.
        p. cm.
    Includes bibliographical references and index.
    ISBN 978-1-4128-1110-1 (alk. paper)
        1. Publishers and publishing--United States. 2. Publishers and publishing--United States--History--20th century. 3. Scholarly publishing. 4. Publishers and publishing--Political aspects. 5. Publishers and publishing--Economic aspects. 6. Publishers and publishing--Technological innovations. 7. Communication in learning and scholarship. 8. Communication and technology. 9. Transaction Publishers. I. Title.

Z471.H67 2010
070.50973--dc22

                                                                                    2010024041

ISBN 13: 978-1-4128-1110-1 (hbk)
ISBN 13: 978-1-138-51384-6 (pbk)

Contents

Dedication

Un homenaje a mis compadres de la generación de los cincuenta en Buenos Aires que todavía siguen con nosotros (Mario Bunge y Lalo Schifrin), y el otro que esta en el gran paraíso (Gino Germani).

Acknowledgements

Chapter 1: Publishing Challenges in the New Century
Originally Published as: "Politics and Publishing in a Democratic Society: Technical Breakthroughs and Research Agendas" (with Mary E. Curtis). *Publishing Research Quarterly*, Vol. 10, No. 3, (Fall 1994), pp. 22-30.

Chapter 2: Technological Rabbits and Communication Turtles
Originally Published as: "Technological Rabbits and Communication Turtles," *Empedocles—European Journal for the Philosophy of Communication*. Vol. 2, Issue 2.

Chapter 3: Tripartite Nature of University Presses
Originally Published as: "The Tripartite Nature of the University Press," *Journal of Scholarly Publishing*. Volume 38, Number 4 (July 2007). Pp. 200-210.

Chapter 4: Limits of Standardization in Scholarship
Originally Published as: "Limits of Standardization in Scholarly Journals" in *Scholarly Publishing*, Volume 18, Number 2 (January 1987) 125-130 pp.

Chapter 5: Publishing, Property, and Information Structures
Originally Published as Two Review Essays: "Publishing, Property, and the National Information Infrastructure." *Publishing Research Quarterly*, Vol. 11, No. 1, Spring 1995, pp. 40-46.

"Intellectual Property and Internet Publishing." *Logos*, Vol. 10, No. 1, Winter 1999, pp. 56-57.

Chapter 6: Specialization in the Electronic World
Originally Published as: "The Assured Future of Specialized Publishers in the Electronic World." *Logos: Journal of the World Book Community*, Vol. 6, No. 3, Autumn 1995, pp. 158-162.

Chapter 7: Social Science and Scholarly Communication
Originally Published as "Social Science and Scholarly Communication in the New Century," *The St. Croix Review.* Vol. 34, No.5, October 2001, pp. 57-60.

Chapter 8: Open Access and Closed Minds
First publication.

Chapter 9: Professional Ambitions and Public Interests
Originally Published as: "Scientific Endeavor, Professional Aims & Public Interests" in *Society*, Vol. 40, No. 1 (November-December 2002) 8-11 pp (With Jonathan B. Imber)

Chapter 10: Formatting Ideology through Tabloid Politics
Originally Published as: "Tabloid Politics: Formatting Ideology." *The St. Croix Review.* Vol. 37, No.1, February 2004, pp. 47-57.

Chapter 11: Scholarly Pornography
Originally Published as: "Scholarly Pornography." *Chronicles*, Vol. 33, No. 4 April 2009, pp. 31-33.

Chapter 12: Publishing Programs and Policy Dilemmas
Originally Published as: "Publishing Programs and Moral Dilemmas." *Journal of Information Ethics*, Vol. 6, No. 1, Spring 1997, pp. 13-21.

Chapter 13: Political Periodicals in Policy Formation
Originally Published as: "Political Periodicals in Policy Formation." *Knowledge, Technology & Policy*, Vol. 11, Nos. 1-2, Spring-Summer 1998, pp. 16-24.

Chapter 14: Monopolization of Publishing and Crisis in Education

Originally Published as: "Monopolization of Publishing and Crisis in Higher Education." *Academy: Bulletin of the American Association of University Professors*, Vol. 73, No. 6 (Nov-Dec) 1987, [Appeared in January 1988] pp. 41-43.

Chapter 15: Publishing Responses to Economic Crisis

Originally Published as: "Scholarly Publishing in 1929 and 2009" in *Culture and Civilization, Volume 1*. Transaction Publishers: New Brunswick, NJ. 2009.

Chapter 16: Publishing as a Vocation: The Necessity of Independence

Originally Published as: "Publishing as a Vocation: The Necessity of Independent Scholarly Media" in *The Journal of Scholarly Publishing*, Vol. 41, No. 2, January 2010, pp. 131-144.

Introduction

Publishing as a Vocation: An Old Profession in a New Technological Era

Publishing as a Vocation is both a companion volume to *Communicating Ideas* and at the same time, an entirely distinct collection that can be read in its own right. The linkage of politics and technology is the driving momentum of our age, and it is assuredly the center of gravity for these studies. It was Friedrich Engels, I believe, who once noted—in a little appreciated revisionist essay—that technology rides roughshod over military systems and profoundly alters economic systems as well. Under the circumstances, it is hard to imagine that the media, of which publishing is a part, would be exempt from so mighty a force.

To have simply tacked on essays to *Communicating Ideas* would have been a disservice to the reader and author as well. For while the world of research and imagination remains what it has perennially been—the moral codebook of right and wrong, strategy and tactics, and ultimately, good and evil—the packaging of publishing, or if you will, its technology, is part of a new epoch. I started working in media studies during a remote age of books, journals, and monographs. The typewriter, that strange mechanical device that is better suited for museum viewing than personal usage, has been displaced by marvels such as voice recognition, and computers that spell and edit. We are now part of the transformation of the slow to the instant. From twitters to bloggers, the communication of written as well as verbal ideas can be transmitted in a matter of minutes now, not years.

To have noted the commonplace should not obviate awareness of the tradition of culture. The continued relevance, even the ferocity, with which issues ranging from the empirical and the ethical continue to inspire the political landscape, is indisputable. But that landscape goes now to the highest bidder, who has the power to persuade and dissuade. I grew up in the era of George Orwell, a journalist and moralist who I greatly admire for his clarity of language and precision of politics. But on one issue I must take strong exception: this new information technology carries with it as much an expression of free and democratic thought as the potential for totalitarian control. Indeed communication, unbound by wires and electricity has given way to the wide expression of masses of humanity, an outcome not always pleasant to the eyes and ears of those who enjoy managing social life. However in this multidimensional equation of production and technology, resides the greatest hope to liberate humankind from ignorance and ideology—and that is what the purpose of communication is and always will be in a democracy—imagination.

If politics is the art of the possible, then technology is the higher art of transforming scientific principles into practices. Together, they comprise and give substance to the title of this book. Theories of politics and practices of publishing form a mosaic of my career. Each person puts together a puzzle that gives meaning to a life. I dearly hope that this volume provides the reader with one more way to do just that—and in the process, gives enlightenment to a process of communicating ideas that remains endlessly fascinating and fulfilling. Beyond such culture written in the logic and language of technology are the yearnings of all past human ages transmitted in pen and paper and all too often in pain and penalty. The scientific world changes rapidly, the moral codes by which we establish norms change far more slowly and devilishly. That is both the message and guide offered in *The Vocation of Publishing.*

As a footnote to such big thoughts is a warning that this is hardly a seamless effort. While most of the essays and lectures herein offered are the product of writings in the twenty-first century, there are sufficient time lapses and intervals between such literary endeavors that advise the reader to see these as independent statements, connected more by sentiments than chapter linkages. In an odd way the volatility of the period is such that even were one to attempt a history or sociology of publishing at this time, it would be threatened and overtaken by events. Under the circumstances it seems best to face up to the fact of the duality of technology and morality, and leave the resolution of present fissures to others.

Irving Louis Horowitz
Princeton, New Jersey
December 28, 2009

Part 1

Technology and Morality in Publishing

1

Publishing Challenges in the New Century

It is best to start with a pronouncement and a belief: publishing flourishes best in a democratic society. This is because there are no external limits, such as state power, on what can be published, by whom, to what ends, and in what numbers. Technological changes alter the relationship of forces so that opportunities for publishing expand even further. A broader range of information channels, especially on the Internet, exponentially increases choice and makes authoritarian rule more difficult. Publishing also enhances democracy when decisions are made on the basis of literary merit rather than top-down political legislation. Those who elect to forget such simple homiletics are soon in other areas of commercial activity.

Research on publishing should reflect awareness of the delicate interaction between publishing and politics. This interaction is reflected in such areas as the relationship between public and private sectors, the impact of internationalization, non-U.S. ownership of information, mergers and acquisitions, and conflict between the First Amendment and copyright law. These are but a few examples of areas in which democratic outcomes are critical and not subject to external controls or limitations—as they are in small places like Cuba and big places like China. The strength of the dictatorships more than the geographies involved determines the ability to impose constraints. Decision-making in publishing would greatly benefit from heightened awareness of the political consequences of publishers' activities. At this point in time, early in 2010, the heads of houses operate in private awareness but public silence on the constraints and opportunities provided by the political system and powers that be.

As the twenty-first century completes its first decade, the relationship of publishing to a democratic society and to human freedom has taken on greater significance than ever. Developments in communications technology, copyright ownership, information access issues, and dramatic international social and systemic changes are converging to have a profound impact on this relationship. Such changes have impacted the very nature of publishing. Its splendid isolation of the past has given way to publishing becoming a small part of the "media" empire, and if it is defined by the political process, it in turn defines that process.

Developments in communication technologies have already had a monumental impact on the educational and workplace environments, and these in turn have affected the public's expectations of publishers. Some of these developments have been written about, but few have thought through their cumulative impact on traditional publishing. For example, it is now possible to work effectively, alone or as part of a group, without being in the same physical location at any time. The Internet offers the ability to share common images and communicate almost as if one were in the same room. Networking makes remote sites simple and effective means of communication. Skype and on-site photographic images, voice mail systems, along with a new generation of "smart" cellular telephones, permit people to establish their own schedules, and to be in contact with others at convenient times and locations. Computers and electronic mail permit instantaneous communication across continents. Portable computers have given way to laptop computers, and these have morphed into handheld devices that combine telephone, Internet communication, and Internet access worldwide. People are able to extract information from large public and private databases. Mini-printers and portable fax machines permit people to print out information received or accessed wherever they are. And mini-scanners enable them to send the images anywhere they want via the Internet. A plethora of print material can be transmitted simultaneously to one central location or to multiple satellite locations. These developments affect not only how publishers work, but what they publish and how they produce it. Indeed, a wide variety of handheld machines that simulate books permit electronic books and journals to be made available through such devices. Habits of the heart are hard to break, so just how pronounced an impact on the general culture these devices will have in the future is still to be determined. I suspect that they will broaden access without much impacting the contents of books, journals, and other organs of public opinion.

It is best to begin by acknowledging that the rapid multiplication of choice in how and when one communicates in the twenty-first century itself represents essential attributes to as well as risks of democracy. As has been well understood in relation to authoritarian regimes in the present epoch, it is harder to control a population through arbitrary and capricious rule when that population has access to a variety of information channels. So rather than bemoan these high-powered but relatively inexpensive "gadgets" as somehow bemoaning difficulties upon publishers, it is far more meaningful to recognize the wider range of opportunities available in democratic political cultures. Such opportunities do not necessarily translate into the higher culture. Many of these channels of ideas and information are displeasing to the same people who shout loudest about the worth of democracy; everything from pornography and demagoguery can be found that raise serious questions about the limits of democracy as simply freedom of choice.

My view is that the impact of these technical devices is less macro than micro in their consequences. That is to say, we see profound changes in how printed images are produced and transmitted to end users, but little transformation of the overall impact or number of such images on the general culture as such. If one may borrow an analogy from an older medium with all its attendant risks, radio, we see publishing as a medium capable of growth in a post-television and coming satellite era, although we also expect only modest changes in the actual content of our publications, just as there has been in radio programming. One no longer looks to radio for docudrama, for example, since television presents such programs in a far more compelling way. But music programming, such as FX and Sirius, with classical and well-known popular items, has exploded on radio. Since music is a language of its own, and its abstract and non-visual components are particularly suited to the highly mobile lives we lead today, we can readily incorporate new technologies into traditional cultures. The same kinds of changes may take place in publishing. For example, while information storage and retrieval might be best suited for formats such as CD-ROM and flash drives, which provides enormous storage capability and flexibility of access, everything from scholarly monographs to mystery novels remains very much the province of the book. Indeed, there is now more variety in book publication than at any other time, and undoubtedly more variety than in any alternative mode of popular entertainment.

This multiplicity of communication media as a characteristic of democratic societies can be seen readily in the enormous expansion of new forms of written communication. We already see a clear preference for electronic mail delivery over regular paper mail service in business-to-business communication, even when the correspondence is not especially urgent, and when the time saved as compared to alternatives is minimal. Even so, electronic mail and facsimile mail machines have not put overnight mail services, private or public, out of business. Documents considered legally important still must travel in an original "hard copy" rather than reproduced format. As changes in communication technology proliferate in the workplace, in private homes, and in educational institutions, they affect the public's expectations as to how rapidly they expect publishers to respond to their needs. This in turn has forced traditional publishers to compete in new ways and with a keen sense of the temporal imperative to communications. Publishers who ignore the implications for what the public expects of them do so at their peril.

Book and journal customers will seek more rapid or random access to what publishers have to offer than present-day production and distribution methods permit. Journals that lack a strong editorial framework, that exist as a periodic compilation of the latest work on a subject, as is characteristic of many important scientific journals, may cease to exist conceptually as a unitary product. Scientists may instead rely on databases to inform them about articles they are interested in, and seek single copies through document delivery services in hard copy formats, or, if available, from their library. The library may continue to receive the print journals through the mail as a subscription for archival purposes. Individual scientists will probably not subscribe, but receive the articles they request through a variety of technologies (Internet/electronic mail, fax, or mail services). The source may be the publisher, a document delivery service, or a library. All of this points to the multi-tracking of information, even more than the heralded multi-tasking of information flows.

We already have audio books and video books, although, like print books, they are still often delivered through existing distribution channels. There has been much discussion of electronic books, which may be delivered through conventional channels (stores, direct mail) or, in the most visionary thinking, from a central electronic library such as Google or Amazon. Such electronic books may offer vastly different possibilities than print, including interactive capability, the ability to extract and manipulate information, and so on. The difficulty here is that many publishers may not believe the audience potential for such books is large enough to warrant the investment of resources required to achieve these capabilities that also risk conventional methods of reaching the public. They may rightly feel, with some justification that many kinds of books would not significantly benefit in quality from such enhancements. The one certainty is that new communication technology will render publisher concepts of what constitutes a book and a journal much more open-ended than at present. The concern will be delivering information, whatever the format, for a decent compensation to publishers and authors alike.

Too many of those who speculate about the impact of communication technology on publishing continue to present the future as a series of alternatives rather than a coexistence of options. In consequence, the economic no less than political implications have been badly misunderstood. Many publishers are deeply concerned about change, in part because publishers understand what they do in a very limited way. They think that they take manuscripts, edit them, put them into type, have them printed, market them, and distribute them to interested purchasers. This parochial vision makes publishers vulnerable to those in the hardware end of the new technology who are alert to developments and who can identify a potential replacement for every one of these specialist activities. Publishers need to confront the sources of their own sense of insecurity before they can understand or explain to others the things they do that are not self-evident and that they may themselves take for granted.

In this area, scholarly publishers play a pivotal, if intermediary, role between the producers and the consumers of our intellectual and cultural life. They encourage scholarly work they believe to be important, either directly (by direct communication with an author) or indirectly (by publicizing their interest in an area of publication). They organize an evaluation process whereby decisions are made about what merits broader attention, involving internal and external participants. Publishers also ensure quality control by verifying that the work is complete and meets objective standards for publication, recommending improvements, and ensuring acceptable quality of production and presentation to a public. They work energetically to bring the author's work to the attention of the broadest possible audience, and ensure that the work is available to those who wish to acquire it. After publication, they protect the integrity of the author's work by ensuring that those who use it or reproduce it acknowledge its origins. They generate and maintain information about who has acquired the work, so they can communicate with them in the future about other work of a similar nature. In this process, inventory maintenance shrinks in comparison to propriety considerations.

Hidden in these new and innovative technologies is an implicit democratizing role. For without universally recognized procedures, the very substance of democratic life is tarnished by idiosyncrasy and arbitrariness. It might well turn out that the need for procedures—what might be called the *methodology* of democracy—will do more to preserve certain forms of publishing than simple technical requirements. An individual working alone may be able to produce a product that looks like a book, and do so at a remarkably high level of competence. What the individual cannot reproduce are the varieties of "value added" activities performed by a decent publisher.

This description of some of the important activities that publishers perform does not imply that publishing as we know it will remain static. Clearly, it already has, and more changes will follow, perhaps of an even more dramatic sort. We now see, through Lightning Source and other like services, not only printing on demand, but binding on demand, thus eliminating the need for large inventories. In themselves, such changes do not threaten the survival of publishing. Quite the contrary, technical flexibility permits far greater emphasis on editorial discretion and marketing outreach than in the past. Furthermore, it is already clear that marketing is taking on a more significant role as a direct exchange between producers of information and recipients of ideas and values. Marketing may even take a greater proportion of publishing budgets than at any time in the past. How publishers will redefine their roles in response to increasing technological change remains an open question. But the old way of reaching a well-defined public only through newsprint is already a mode of the past.

Another area of publishing that merits the attention of those interested in the democratic society is in part related to the above, but is sufficiently distinct as to require independent assessment. Some developments in communication technologies may have a dramatic impact on the relationship between the private and the public sectors in publishing and publishing-related activities. The federal government generates large masses of information as a by-product of its activities. It also funds significant areas of research and development. Until recently, public dissemination of this work has generally been limited to print formats, and up to now it has by and large been ineffectual. The private sector has stepped into this vacuum, playing a principal role in retooling, packaging, and selling government-generated information. Thus publishing, scholarly and reference publishing in particular, has become a forum in which issues of publicity and privacy are both debated. Such debates reflect upon the sensitive area of the limits of control by government at the top, and what can legitimately be withheld by individuals at the bottom. The census data are a perfect illustration of where these two ideals meet, and not always in comfort.

Many participants in publication of government-generated information or government-financed research are members of groups such as the Association of American Publishers and the Information Industry Association. Many are major publishers of traditional books, magazines, and newspapers, influential sectors of the American public, and librarians in particular, have voiced a belief that these companies have no business reselling to the public what it has already funded through taxes paid. In addition, some of these publishers are not American-owned companies, although they may have been at one time. Their repackaging and reselling of government-generated information, and their publication of research funded by the government, has aroused concerns akin to protectionist demands being voiced in other economic sectors. Some critics have even raised national security concerns, arguing that it is shortsighted to permit control of critical segments of American information resources to be in the hands of non-U.S. companies.

The latest and most militant development is the demand for "open access" of all materials produced in semi-publishing settings, such as universities, agencies, and foundations. The argument by such advocates is that publishers do not participate in the costs of such information, and therefore should not benefit from an exclusivity of copyright and publication. Again, this is a demand for more, not less, democratization. Whether cutting out the publisher is feasible, given the costs of securing editorial advisors, and paying the freight for a variety of marketing and warehousing features, is even possible, is beyond the scope of this opening chapter. But it is evident that pressure by academics no less than librarians has moved the discourse to a sharper vision of identifying democracy with open access of information.

This is not the place to argue the premises underlying such policies. However, it is reasonable to say that avoiding monopolies on information sources and delivery is itself a function of the democratic society. The worry that a few may profit on occasion from repackaging publicly generated datasets is more than offset by larger concerns: the consequence of government functioning as a monopoly on access to and use of information. Government-generated information is exempt from copyright constraints—in part because the Founding Fathers appreciated the need for public access to discourse on central considerations of policy. How the private enterprise sector in commercial and professional publishing alike respond to such urgings will probably be worked out in court rulings and decision, which in turn must secure a broader outreach of materials and at lower costs, and yet insure the viability of publishing as a mechanism for decision making on the quality of performance no less than the costs of publication as such. In this sense, the public-private partnership which currently exists is itself part and parcel of the democratic approach to running government.

Added to the equation is government support of a precursor to a high-speed electronic communication highway, the Internet. As the Internet has become a worldwide phenomenon, and attracts more and more users, some have challenged traditional norms of behavior, from publisher control of copyrights to payment for products. Those who are committed to a free-market, private-sector approach to scholarly publishing are being challenged by those who believe in public-sector approaches; this debate is emerging as a version of the old "right to know" versus "right to ownership" arguments that characterized the mid-eighties. It is hard to imagine Bitnet challenging newspapers in the delivery of information and entertainment to large numbers of people. Yet it is also hard to deny that Bitnet does greatly amplify and enhance news reports. For example, a recent article in the *International Herald Tribune* concerned charges that an ex-student, had been charged with spying. The *Herald Tribune* story was a one-paragraph capsule, but Bitnet carried two full pages on the same story, gathered from a variety of press services such as Associated Press and Reuters. It even included direct interviews of subjects who know the accused individual. Again, the relation of democracy to publishing is not a simple linear equation, but a complex set of public choices, the depth and detail of which are determined by the exact need to know specific things at specific times. As a result, both delivery and acquisition of information and ideas are an interaction, not just a dissemination. This empirical truism is the common meeting ground of democracy and publishing. It is also the flashpoint where gossip and slander interact with information and opinion making.

Since much of this discussion has focused on new opportunities, perhaps it is also important to take note of new limitations on democracy within the publishing world. One such limiting element in the debate over democracy is the increasing internationalization of publishing, coupled with monopolization of the industry on an unprecedented scale.

In the last few years we have seen the merger of Pergamon with Elsevier, Reed with Cahner, and now Reed/Cahner with Pergamon/Elsevier. New arrangements between Wiley and Blackwell, Taylor & Francis and Informa have served to further concentrate market share. One may claim that there remains a plethora of publishers about, but that disguises a higher truth, the monopolization of the industry in many sectors. It might well be argued that five large journal publishers control close to 90 percent of journal publications; with smaller university press ownership has dwindled. Monopolies are themselves in constant struggle with each other. Closer to home we have Macmillan becoming Maxwell-Macmillan, then Paramount buying Macmillan and merging it with Simon & Schuster and a myriad of sister publishers such as Prentice-Hall. In turn, Paramount has been absorbed in the acquisition contest by Viacom. Previous acquisitions of publishers by publishers tended to result from interest in developing broader publishing capabilities: trade publishers acquired a scholarly or a textbook publishing arm, for example. Even in a previous wave of mergers and acquisitions in the 1970s, non-publishers that acquired publishers tended to be conglomerates interested in developing a broad-based portfolio of companies in different economic sectors. They were not primarily concerned with developing a commanding monopoly presence in any one industry.

The acquisitions of the twenty-first century have accelerated this process. Whole sectors of publishing have come under the monopoly control of single giant companies, and these are often multinationals not based in the United States and hence not subject to U.S. antimonopoly legislation. The U.S. government has tended to turn a blind eye to mergers that in earlier epochs might have been prevented as monopolistic, perhaps aware that by enforcing such legislation, they had put U.S. companies at a competitive disadvantage to non-U.S. companies. The consequences have been unsettling. What does it mean when one of the two or three largest elementary textbook publishers in the United States is owned by a non-U.S. company? What does it mean when the world's most important scientific journals are published by a single (non-U.S.) company? The challenges to those who are committed to the conscious promotion of democracy, or for the less abstractly inclined, to private-sector publishing, are enormous. And the implications for the historic relationship between publishing and democracy are profound. Once again, this is intended as a prolegomenon to a very big subject, not a definitive, one-time answer.

Other developments in publishing could well benefit from analysis that considered the implications for democratic outcomes. The brewing confrontation between the First Amendment requirement of open access and copyright law is one such concern. In a number of areas our historic constitutional protections of free speech and privacy are being challenged by those who argue that copyrights are being violated. Publishers have strong commitments to both traditions. The Association of American Publishers has several core committees: Copyright is one, and the Freedom to Read (First Amendment) committee is another. In a number of recent instances, these interests have sharply diverged. For the most part such differences have been papered over through legal rulings that toss a bone in either direction, but they are unlikely to resolve the larger considerations in a definitive or at least unambiguous fashion.

One older judicial ruling that still haunts the publishing community is the case of HarperCollins (then Harper & Row) versus the *Nation* magazine; Harper sued the *Nation* for publishing extensive prepublication excerpts from the memoirs of former president Gerald Ford. Harper had sold prepublication rights to excerpt the book to *Time* magazine for a substantial sum; *Time* canceled the agreement, claiming the *Nation* had preempted their publication. The *Nation* claimed it was simply publishing the excerpts as news; it had obtained the information through an undisclosed third party. The Supreme Court decided in favor of Harper, but within a narrow band of judicial opinion—one that recognized the public's right to know about political leaders. We see evidence of increasing tension between publishers, especially publishers of books in contrast to publishers of magazines and newsletters. These are part and parcel of the proliferation process itself and, one must add, of the democratizing process as well. The tendency is for such copyright infringement cases, especially those in which actual proprietary ownership is hard to establish, being settled out of court and in a routine fashion.

Illustrating these inner tensions are micro issues related to proprietary rights. In a number of instances, authors or their heirs have challenged the use of letters and other unpublished materials in scholarly works. Because some of these were on deposit in libraries, librarians have also been concerned. Such uses were open to challenge because the Copyright Law of 1978 states that copyright exists from the moment of creation, not from the moment of publication. Such challenges have in turn created problems with use of such unpublished materials in biographical and other works. Authors or their heirs claimed violation of copyright, and they were victorious in several important legal suits. Looking to the *Nation* precedent, the courts asserted there could be no "fair use" of unpublished materials. First Amendment advocates envisioned a disastrous situation in which entire areas of scholarly research were effectively cordoned off.

For the time being, the tension has been resolved by legislation extending the concept of "fair use" to unpublished materials or materials where the original owner is difficult or impossible to locate. Once the Association of American Publishers developed a concerted position on the need for this legislation—no small task given the inherent tensions between competing interests within the organization—they were faced with the task of convincing other interested parties to share in this restraint on information usage. In particular, the computer software industry was opposed to such an approach, feeling that "fair use" of unpublished materials would weaken protection of their properties. Since billions of dollars are lost to the industry each year through illegal copying of software, their concerns are understandable. However, a temporary consensus, more like a truce, was worked out; all sectors of the copyright industries were assuaged; and the legislation was successfully put forward. Democracy is, after all, not so much a thing as a process, and one must expect publishing to be an area of struggle in the redesignation of democracy, and not just a residual source of information and entertainment.

This is not necessarily the end of such concerns. An author has recently challenged the publication of extensive excerpts of a letter in an article in *Harper's* magazine. The case was heard by the same judge as in the Texaco suit, and the author won, although the award was small. Heirs have challenged use of letters in biographies of which they do not approve. Others have sought to control a vast array of digitized material through special arrangements with university archives. It is apparent that copyright law is being used in attempts to suppress publication at one end, while these same laws are aimed at breaking the back of property concerns as such. These are both dangerous uses of copyright.

Another dangerous trend is insensitivity on the part of publishers to the balance of rights inherent in copyright law. This takes the form of vigorous emphasis on the property-right component in copyright by publishers and inadequate attention to dissemination. The Texaco decision may have moved the law in the direction of proprietors, but more recent exemptions indicate that they would be well advised to be careful. In that suit, Judge Harold Lasker was careful to note that Texaco should pay for the documents it had photocopied because it would not be an undue burden for them to do so. Now, there are two components of "undue burden." One is reasonable access; the other is reasonable price. If publishers are inclined to try to obstruct the development and application of new technologies, they may find the law to be less responsive to their position on copyright.

A measure of the importance of research on publishing is evident in the launch of several important new journals specifically devoted to publishing in recent years. *Scholarly Publishing* has existed for more than twenty years; but we now have *Publishing Research Quarterly* (formerly *Book Research Quarterly*) and *Logos,* as well as *Publishing History,* along with any number of trade magazines. In addition, there are a variety of newsletters and other less formal media of communication. These publications do not exist because people need to write in order to secure their position in the world of academia. There are few tangible rewards for publications in the world of business (of which publishing is very much a part), and it takes time away from tasks for which people are held strictly accountable. This has changed, in part, as publishers have seen the benefit of having a body of literature presenting their perspectives on issues of importance to their business. That said, even such specialist journals in publishing at times fail. For they too are faced with the costs of hard copy versions in a university of users that demand rapid as well as easy access via electronic sources—but not quite willing to pay for such accelerated deliveries.

In this sense, what is needed is far more extensive training on the part of publishing executives in the political system as well as in the editorial, sales, and marketing areas. It may well be that courses in political science may prove of greater value that those in business management. For those who serve the liberal arts as well as the sciences in a publishing context, it is essential, not optional, to have a sense of the needs of the political world, the character of the political world, and the ethics (or lack thereof) of the political world.

We suspect that the relationship of publishing to democracy carries over into an appropriate management style within the publishing environment. The very process of refereeing and reviewing, of selecting among alternative typefaces and designs, indeed of approving what to publish and what to waive, all involve the intimacies of democracy. One comes face to face with the fact that—within limits—contemporary forms of publishing, especially in professional, technical and scientific work, involve decisions of many people based on competence and skill rather than authority and rank. In that sense, the developments we have spoken of in the larger society have profound ramifications for different styles of management at different levels of decision-making within distinctive publishing contexts. A book is a unique product, even if it has in common with comparable books, shared problems. That is why publishing is an industry with bulk and weight, but one that cannot yet be measured by the pound.

We started with a broad pronouncement on the relationship of democracy and publishing, and the legitimacy, even the centrality, of such concerns. To focus specifically on this theme in a new technological context is to take seriously changing definitions of what belongs to the public square and in equal measure what rights inhere in the private person. For the goal of the word whether in hard copy or electronic form is to enhance the access of individuals to the world of the rich and famous and no less, the wicked and infamous, not to mention everyone caught in between these polarities. At some point, we are star-crossed between a definition of the democratic as an information supply for the masses, and a rather different definition of the democratic as the right of the individual to remain silent, and to abstain from the rush to judgment. The enormous supply of information makes this a new issue indeed, or better, a new bundle of issues. It might be best to disaggregate them if we are to cope with a publishing environment that seeks to remain true to the highest ideals of the democratic society without risking a further erosion of variety and options that permit pluralism of goods and services. In short, democracy is a double-edged sword: a better informed public and a more secure publishing base.

2

Technological Rabbits and Communication Turtles

In the first class of the first course that I took in philosophy at the City College of the City University in New York, the instructor, a fine scholar by the name of Mortimer Kadish, who went on to a distinguished career at Case Western Reserve University, asked a seemingly harmless question: "Why do we look upon the mathematics of Archimedes as a curious artifact, and the ethics of Aristotle as an essential text in the learning process of our age?" I recollect that no one in that class was able to understand the epistemological significance, much less answer that question! I am not certain that I can resolve this paradox even at this point in time, but at least I now know the burden of such a question.

What prompts such a stark recollection to a dim past is a review of a publication entitled *Technology First: Journal IV* with the inevitable theme of "The Next Step in Communication." The issue features articles on consumer demand for cheaper, easier broadband access, with Wi-max or the tongue twister: worldwide interoperability for microwave access. This is followed by a piece on the greater use of flame retardants to reduce fires and injuries that come from electrical faults. It emphasizes the use of HIPS or high-impact polystyrene and Acetonitriler Butadiiene Styrene. This is followed by information on the latest devices to test functionality and connectivity of the entire UBS protocol on a single chip. A follow-up essay discusses how, in an age in which 90 percent of people in the United Kingdom have a cell phone and

50 percent have Internet access, the issues sharply change from communication between people, with devices, and even how devices communicate with each other, to the growth of wireless connectivity as an end unto itself—thus replacing cables and the need to increase security.

The pattern of such empirical research is now well established and doubtless known to most of the people working in the field of information technology and communication. There is little point in reciting all of the amazing developments taking place that are truly changing the shape of the physical and human environment. It is evident that technical applications are the critical factor reshaping the character of our lives: from how we read, what we feel, and when we perform certain routine and extraordinary actions. Entire industries are dissolving before our eyes, from newspaper closings to automobile collapses. At the same time, and in reverse, entire industries are coming into existence: from open access to information in a variety of electronic forms to automobiles driven by miniaturized battery devices that threaten to make petroleum an obsolete fuel. Everything from the nature of education to the power of extending human life through medical devices and drug regimens is at stake in such changes.

In the decision-making process, communications experts and social scientists provide different qualitative inputs. The former provide expertise which tends toward non-ideological, non-political solutions to social problems. Systems design displaces ideological disposition for many in the field. Social scientists, for their part, tend toward political solutions. The latter are usually wary of technological bureaucracies sponsored by governments as technicians are of charismatic or quasi-patriarchal domination by governments.

This divide between communication research and social scientific studies was not always the case. For many years, especially in the middle of the last century, communication studies were very much a part of sociology and to a lesser extent political science. People like George Gerbner, Ithiel de Sola Pool, Elie Abel, Hans Speier, Harold D. Lasswell and Hugh Dalziel Duncan, to name a few, participated greatly in the sociological imagination. As the field of communication studies advanced its own academic and research agenda, it developed strong empirical tendencies, derived in some measure from public opinion survey research and propaganda analysis across boundaries. Such a movement toward the quantitative was stimulated if not entirely supported by giant corporations in the field of communications. In short, a field that hardly existed at the turn of the twentieth century took on special service characteristics that moved it away from social science as a source of intellectual energy.

Connections to older social sciences were seriously weakened over time. The examination of moral issues has just about vanished among no-nonsense technicians. Communication studies became a field unto itself, with strong statistical and quantitative guidelines that set strong boundaries to what was considered viable or valuable. These parameters became the measure of professionalism, and one would have to say that in numerical terms this great divide allowed for the growth of communication studies to a point that it has dwarfed those areas from which it emanated. This is hardly an unprecedented phenomenon.

With success come new problems. Moving away from one's own history is a mixed blessing. The problem can be stated as follows: the communication impulse is toward rational, systematic solutions, and toward impatience with the norms and values of non-rational and ideological, class, radical, and ethnic interests. Like most professional groups, people doing research on communication have been reticent to perceive themselves as yet another special interest group whose claims to political pre-eminence are neither more nor less valid than the claims of other interest groups. The communication field sees itself as largely a professional group, part of the great expansion of technocracy no less than technology. Indeed, with a glacial shift from the means of production to the means of communications, and specifically from measuring wealth and power in terms of proprietary rights rather than inventory of products, the impulse toward technocracy becomes sharper, and assuredly more transparent.

The most vital issue immediately affecting technologists and social scientists alike is their joint commitment to developmental application, to framing a language amid a set of techniques widely understandable because they rest on results based upon commonly employed systems and designs. Thus far, the scientific language of system design and manpower allocation has not been fully integrated with a political language of social transformation and decision making. This is not to fault either professional group. Indeed, social scientists have reinforced present methodological divisions by assuming stylistic differences between systems management and applied research to be permanent simply because they are operational. The emergence of computer technology, widely adapted to social science uses, and the corresponding emphasis on measurable indicators of health, welfare, fertility, crime, etc. has clearly linked the two groups so as to make new breakthroughs and cooperation highly probable in the next stage of communication studies. But these breakthroughs are linked to practical applications, such as security concerns and privacy issues in law and behavior.

Any serious dialogue between social scientists and communication experts requires a recognition of the enormous contributions made by modern technology to social science such as: (a) the engineering emphasis on the special problems involved in applications and direct research, which has created a whole new field of social planning; (b) the managerial emphasis on posing problems in soluble formulations, rejecting apocalyptic readings of events as a way of bringing about a solution to problems of the world; (c) the emphasis on design, delivery ,and organization, particularly with respect to the processes of urban expansion and industrialization. This in turn has permitted social scientists to place greater emphasis on anticipating problems of social reorganization by conceiving the larger society as appropriate for the input of parallel design, delivery, and organization. The devil in this bargain is the loss of interest in larger ethical domains—both for mass communication and social research.

The reference to a link between humanities issues and the social sciences has itself come to be viewed as an indicator of parochialism. Technicians often think of social studies apart from an understanding of social sciences. Further they tend to view social science inputs as shadow rather than substance. For example, many important writers see social problems strictly in terms that can be treated by technicians, for example, environmental pollution, transportation, water resources, medical care, regional development, and so on. No mention is made of problems of revolution, terrorism, anomie, generational conflict, racial struggle, etc. The view of the communications personnel remains insular and cautious to a fault. It is clear that as long as commonplace pieties are used to describe the social sciences, the contributions of the history of social science to communications research, with specific reference to developmental analysis, will remain peripheral, largely unexamined, and perhaps an irritant.

There is a strong tendency on the part of information technicians to celebrate their intellectual self-sufficiency. They now speak of a doubling of knowledge in decades. They refer with pride to the fact that a history of technology covering two earlier centuries may take one volume, while the same sort of historical survey of the twentieth century alone may be a multi-volume affair. They even manage to get some scholars to promote the heady wine of futurology, or "one hundred technical innovations likely in the next thirty-three years." The more judicious engineers speak in sober and cautious terms, pointing to the relative slowdown in new inventions in recent years, and the under-utilization of present plant capacities. The less cautious among them speak of how robot technology displaces human will in the execution of the tasks of armed conflict and open warfare.

What has taken place, and what obscures the gulf between disciplines is the sophisticated refinement of established inventions, or more precisely a lag between invention and applications. In the first half of the twentieth century, major inventions were realized and widely applied: plastics, automobiles, radios, television, and commercial aircraft. In the second half, one may list electronic computers, wireless transmissions, video tape, nuclear explosives, robotics, and antibiotics. We may now be entering a period of invention refinements rather than breakthroughs. If there has been a slowdown, as a sober reckoning would in fact indicate, and if we are getting a phase of application in place of innovation, then it is important that social dimensions of the problem of innovation be raised and celebrations be modified.

The shift in emphasis from the means of production to the instruments of communication is highlighted by the fact that communication as such changes the character, design, and structure of production as such. The development of wireless communication illustrates this shift. In the airline industry passengers can now use cell phones as two dimensional bar codes that can be displayed and scanned at boarding gates and security checkpoints. In the auto industry, all phones can now start the car and adjust the seats; digital mobile keys are in the making. Bigger screens and better browsers are now allowing smart phone users to do online banking. In education, the mobile phone is equivalent to the laptop, and hence an instrument for basic changes in pedagogic techniques. In energy, the incorporation of smart meters now permits home owners to control energy usage and consumption. Medical records are increasingly incorporating mobile devices to maintain records, adjust drug usages, and text message patients directly. Mobile web bookings have become commonplace in the hotel industry. Political campaigns increasingly rely upon text messaging with supporters and to recruit volunteers in mobilizing for critical campaigns. Finally, closer to home, publishing has witnessed the rise of a new era of digital, mobile users that permit wireless data modems that change procedures for everything from pricing production to reading classics.

An impressive fact is that new inventions and applications are increasing at so great a speed that they cannot be readily absorbed at quotidian human levels. More often than not, they require collective efforts that far transcend the capacities of any single person, or sometimes even an industry. No less significantly, the sort of inventions now being brought out cannot be counted on to always produce rapid returns on financial investments. Hence, private firms are reluctant to invest in large-scale research, preferring refinements of presently marketed items. Public-sector enterprises, for their part, are hamstrung by the lack of funds, or by constant pressures within certain societies against extended investments in new projects that may have little chance of commercial success.

Because of this gap between economy and technology, at least in the area of commodities, the gap between invention and wide-scale usage must be viewed as structural and long-range in character, rather than representing a temporary lag in the overall economy.

Instead of coming to terms with such essential socio-structural issues, what Robert Boguslaw termed the new utopians—those technicians turned social analysts—have let loose an exaggerated burst of optimistic appraisals about the social structure. The shift in technocratic orientations is away from the social role of scientific specialists, toward a belief that political or administrative responsibility may become a relatively minor prize in the environment game. In a new view of life the technological vision of social organization has gone along with utopianism. The task of the human engineering principle will be in matching the constructed environment to the person rather than vice versa.

To date, this relation of the person to the machine has been a mismatch. An intriguing piece on "Faux Friendship" by William Deresiewicz, gets to the heart of the dilemma: "Actual human contact rendered 'unusual' and weighed by the values of a systems engineer. We have given our hearts to machines, and now we are turning into machines. The face of friendship in the new century." This claim turns the French Enlightenment theorists who argued for "man as a machine" on its head: "Humanism"—that much maligned word, is now seen by some as a struggle *against* the machine. At one level, the problem resides in the metaphor. Yes, human beings are machines. Then again, they are also animals. But it is clear the whole is far greater than these parts, or if you will, metaphors.

Behind this variety of reductionistic scientific vision are long standing Platonist echoes of the military and intellectual roads to power. Bluntly, we are never told how those doing the "matching" of technology and morality will be chosen. In keeping with this utopian republic underscored by mechanical Power Point displays is a downgrading of entrepreneurial or human administrative skill. Multiple computer uses, and now wireless electronics, make it commonplace to shift the source of error to machines rather than to men and women. In the brave new technological complex, leadership will be done away with, since in a self-sufficient, self-confident, highly educated democracy, all one wants are results. The continued emphasis within the social sciences to ideological explanations is not only dangerous unto itself, but postpones the inevitable coming together of technological and sociological levels of analysis as a means to some larger ethical purposes.

These various developments have led to a new definition of literacy; or computer literacy. This in turn has led to a huge gap between haves and have-nots in the way of modern society as such. Entire segments of society, often those in poorer, working class, and minority situations, are not able to understand, much less utilize the revolution in communications. As a result, any notion of egalitarianism gives way to new forms of social divisions, often with potent consequences. Platonic echoes become pandemic, an acceptable way of distinguishing the gold people from the bronze people. It is risky to speak of education as the mechanism for overcoming such distinctions, especially in a world in which not every person sees education as a panacea to a proper way of life. Indeed, the issue of what constitutes a free society itself is subject to re-examination in the light of new developments. At present, the concept of freedom has been dangerously reduced to information processing.

This brings us full circle to the nature of this contribution to ethical concerns. In the midst of such volcanic change, what is permanent? What is durable? What is even worth preserving? What is the legacy bequeathed to the present? In a world that has elected presidents on the theme of change, what do we do with the permanent things? Indeed, are there such things, or is simply a clinging to the past, a vain expectation that our inheritance itself is something more than an artifact? Whatever the answer is, it is clear that it will not be found exclusively in technology first, last or always! It is my hope that at least some sensitivity to such matters can be at least a small part of the rush to the future.

In an essay on "Engineering and Sociology" some forty years ago, I expressed my debt to Max Weber by differentiating the technological from the social, the mechanical from the human. But it was then an unresolved framework. Only belatedly did I discover the great work of the autodidact Lewis Mumford, and of the geographer Jean Gottmann. It was these outsiders who added red meat to the dry bones of communication theorizing. All of these figures share a common debt to a more distant past. In considerable measure the problem of the disjunction between the technical and the moral resides in an underestimation of that great figure in the history of philosophy, Immanuel Kant. He was the primary source of both Weber's thinking and that of Hannah Arendt. He uniquely appreciated the fact that while change is part and parcel of thought, the rates and styles of such thinking are by no means moving in lock step. This brief set of remarks is a means to correct that oversight.

From Aristotle to Kant we learn of the dangers of dogmatism, of a vision of the scientific and religious orders that proceed without taking into account the sturdiness of quotidian ethical guidelines. The weakness of my earlier formulations stemmed from trying to assign to sociology a level of specific that could correct the abstract nature of technological change. The social is not yet the normative. That is reserved for the Kantian transcendental a priori, the nominal world that is the ghost in the machine of the phenomenal world in which communication travels.

There is a second line that runs from Kant to Arendt. In her remarkable trio of studies on the *Life of the Mind*, she reminds us that in the faculty of judging, as Kant indicated, there is a gap between thinking and judging, between consciousness and conscience. I believe it is fair to say that this extends to the difference between technology and philosophy. This is not about fatuous dualisms, or unmanageable gulfs, but a recognition that all specific fields of endeavor answer to ethical judgment. Arendt summarizes the point well: "The manifestation of the wind of thought is not knowledge, it is the ability to tell right from wrong, beautiful from ugly. And this, at the rare moment when the stakes are on the table, may indeed prevent catastrophes, at least for the self." Such a view does not destroy the grounds for discovery and innovation, but it does place such technical feats in a broader context of the history of human thought.

The grounds for a renewed integration of diverse fields stems from a recognition that although the permanent is the dialectical opposite of the transient, or more bluntly, the eternal is the converse of the changing, both have a place in the normative scheme of things. It is well enough to speak of the inevitability of change—either in terms of improvement or deterioration. And while that is certainly the case, the inverse of that commonplace, and the eternality of permanence is more difficult to accept. The notion of permanence involves a metaphysical presumption of durability, things or ideas beyond those which are subject to quantitative or empirical improvement in demonstrable doses. The strength of the Kantian categories derives not from assumptions about teleological judgment or transcendental mandates, but rather it is based in the long tradition of mathematical suppositions about the axiomatic nature of numbers, and the admittedly difficult assumption that these can somehow be translated into useful moral categories of right and wrong, good and evil, truth and beauty. We can operationally divide these, dismiss some of them, but they have a peculiar ability to pop up in the strangest places.

The cases of abortion and stem cell research are good indicators of how ethical norms change over time. The former is accepted as a contraceptive device of last resort, the latter is understood to be highly risky but a necessary genetic alteration that possibly may save lives. But such procedures are now widely accepted nonetheless. So too are the multiple uses of the cellular phone: to keep parents and their offspring in touch with each other, and also to permit the worst criminal and murderous elements in society to avoid capture! The political realm in advanced nations does not celebrate several million cases of abortion annually nor does society adopt mass eugenics as a strategy, but it dares not use risk factors in innovation to deny the need for legal safeguards to unwed mothers, innocent citizens, or the restorative capacity to the damaged limbs of children.

The point of this exercise is to note that moral order does exist and changes exist in its structure. Clearly, rooted values do not change with the volcanic rapidity of developments in the technological order. So we are faced not so much with an oppositional framework of the technical and the ethical, but rather measuring different rates of change, different speeds at which change take place. This is an important lesson in democracy, since it allows for civility in interaction, and liberality in framing rules for exchanges between systems as well as individuals. The great past wars of science versus religion and theology stem from a belief that they are diametrically at odds—with the assumption that science represents freedom and religion dogmatism. But a close look at the utilitarian nature of computers would doubtless indicate that they are as widely used by clerics as by scientists.

Traditional habits of the mind dissolve slowly, but they do indeed deserve to be reviewed and revised in the twenty-first century—not by dismissive mandates or government edicts; so much as a recognition that changes occur at different rates in different spheres of life. Some changes augur improvement; others seem to follow laws of entropy and decay. The task of the social sciences is measurement not so much of the worth of invention, but rather of the reasons why these rates vary substantially between the technical and the social. The vocation of publishing records such changes. If this is so, what are the variant relations between areas of human life? Leaders in communication have a golden opportunity to serve as bellwethers for this new paradigm because they have a foot in both the technical and social sides of the contemporary system. In this way the dogmas of the past will be buried, and the promise of the present, could become a source for future civilities. This in turn can lead to a greater appreciation of the democratic imagination, at least greater than then I understood so imperfectly in the past.

The notion of normative behavior resides in some measure on assumptions beyond the sociological or the anthropological. This is not to say that technical innovations do not automatically modify assumptions of the moral order. It is rather to say that human beings review such innovations slowly, haltingly, with political resistance and at times, personal resentment. Humans adapt to change on a need only basis. Within the bowels of scientific interpretation and constructive achievement are possibilities of terror—massive destruction on a scale that is unparallel. Five hundred years of Western Civilization should make it clear that change is no magic bullet for progress, nor for that matter are new formulas for social order policy mechanisms guarantees of such progress. We are left with decision making in which mistakes in judgment are more lethal and total than in the past, so caution is the handmaiden of decent scientific method. It turns out that ethics is the durable ghost in the technical machine, but also limits the social whims of policy making and human engineering as well.

These remarks are not intended as a frontal assault against postmodernity and certainly not a critique of new information and communication technologies. The sort of Luddism that hides behind moral absolutes as a means to lay at the foot of technology everything from identity theft to medial ailments that result from the use of cell phones has been repudiated too often to require further elaboration. What does need to be emphasized is how the very shift in technological models of growth can hugely impact everyday events.

The speed with which revolutions and rebellions, regicides, and genocides are reported change the character and possible outcomes of a specific event. Instant reportage does not create the need for such measures as people will undertake in the name of democracy or in defiance of dictatorship, but they offer prospects of rapid moral judgment that tear away the obfuscation of those seeking to work and dominate behind closed doors. In short, while George Orwell's "Big Brother" watching the actions of ordinary citizens with the aid of advanced technologies does exist, so too does this technology just as readily turn its gaze on the big brothers. Communist officials in China order Google to cease providing access to overseas access in the Chinese language, in turn, Google is considering abandoning the Chinese market rather than submit to such political dictates. At the same time, the Iranian regime did everything possible to block access and usage to all Internet services during the post-election riots of June 2009, but to no avail. The information was recorded, photographed, and sent forth worldwide. In commenting on the mass uprisings against the clerical dictatorship in Iran, Peggy Noonan caught the spirit of how moral judgments impact technical discourse for the purpose of political change. Her remarks merit close attention: "The great question is what modern technology can do not in the short term as the long. It is not the friend of entrenched tyranny."

Thus, the answer to the opening query of Mortimer Kadish was implicit in the very context in which it was asked. It is one that I believe he would have wished us to acknowledge and appreciate. Yes, change is different in kind, space and substance for technology and society alike. But the history of ideas is not vanquished by executive decrees. This is why the irritating paradoxes of philosophy remain intact—as a useful preventative for correcting the hubristic impulses that infect the educated classes—in technology, politics and society alike. These hugely differential rates of change leave a space and a place for the humanistic purposes of the four inherited fields of philosophy: epistemology, logic, aesthetics, and above all, ethics—power points in their own right. Such differences in the rates of change in human life make possible better decisions in social policy and technical goals. Synthesis may remain a goal, but one of far less worth than dogmatic assertions from all quarters as to systems and even to the unity of science.

The plain and simple fact is that public policy is best when it navigates between technology and morality, and in recognizing that rabbits and turtles both have the right to live in the animal kingdom. When policy dramatically veers to one side or the other, the results are not synthesis, but chaos. In a democratic society, the task of policymakers and intellectual guardians alike is to make sure that these delicate balances of social forces are respected. The future will attend to resolutions of this grand sort.

References

Hannah Arendt, *The Life of the Mind, Volume One: Thinking*. See the segment on "Banality and Conscience" in *The Portable Hannah Arendt*, edited by Peter Baehr. (New York and London: Penguin Books, 2000), p. 414.

Robert Boguslaw, *The New Utopians: A Study of System Design and Social Change* (Englewood Cliffs, NJ: Prentice Hall, 1965), 213 pp.

William Deresiewicz, "Faux Friendship," *The Chronicle Review/The Chronicle of Higher Education* (December 11, 2009, Section B): 6-10.

Hugh Dalziel Duncan, *Communication and Social Order* (New York: Bedminster Press, 1962), 475 pp.

George Gerbner, *The Analysis of Communication Content; Developments in Scientific Theories and Computer Techniques* (New York: John Wiley, 1969), 616 pp.

Jean Gottmann, *Megalopolis: The Urbanized Northeastern Seaboard of the United States* (Cambridge, Massachusetts: MIT Press, 1961), 810 pp.

Irving Louis Horowitz, "Engineering and Sociological Perspectives on Development: Interdisciplinary Constraints in Social Forecasting," *International Social Science Journal* (UNESCO), (Volume XXI, Number 4, 1969): 545-556.

James E. Katz, "Values, Technology, and Administration: The Weberian Inheritance," in *The Democratic Imagination: Dialogues with the Work of Irving Louis Horowitz*, ed. by Ray C. Rist. (New Brunswick, NJ and London, Transaction Publishers, 1994), 9-38.

Harold D. Lasswell, *Democracy Through Public Opinion* (Menasha, Wisconsin: Banta Publishing Company, 1941), 176 pp.

Lewis Mumford, *The Myth of the Machine: Technics and Human Development* (New York: Harcourt, Brace & World, 1966), 342 pp.

Peggy Noonan, "Whose Side Are We On? You Have to Ask?" *Wall Street Journal* (Opinion) (June 20-21, 2009): A13.

Ithiel de Sola Pool, *Forecasting the Telephone a Retrospective Technology Assessment* (Norwood, NJ: Ablex Publishers, 1983), 162 pp.

David Shen (editor), "The Next Step in Communication," *Technology First: Journal IV* (2009): 32 pp.

Hans Speier, *The Truth in Hell and Other Essays on Politics and Culture, 1935-1987* (New York: Oxford University Press, 1989), 358 pp.

3

Tripartite Nature of University Presses

It is a risky enterprise for a publisher revisiting his older state-ments to presume the role of prophet or seer. To start with, why remarks made over thirty years ago should still be considered relevant requires explanation. It would be incorrect to presume an equivalence of conditions under which such remarks were first made, much less how the comments illumine changes and conti-nuities in the current world of scholarly publication. I should state frankly that I firmly believe that the tripartite nature of publishing still exists—and has remained so for much of the history of that special world of professional communication. In a nutshell, the means by which information is delivered has expanded markedly beyond the printed page. By adopting modern technologies such as information websites, subscription databases, online help centers, CDs and DVDs, scholarly publishing has become part of a com-munication network. That said, the essential purposes of scholarly publishing have remained essentially the same not only for three decades but for three hundred years.

The feudal or medieval element has grown stronger as publica-tion has become an even more critical point in the evaluation of appointments and promotions for university personnel. Universi-ties grow larger and hence need to assign the role of decision-mak-ing to surrogates such as press directors and specialized editors. At the same time, the scholarly community participates in the highs and lows of university and institutional funding. Library budgets in particular have become a critical element in determining the success or failure, and even the life and death of scholarly publications. As a consequence, the struggle for the market by marketing personnel has become a core element in the business of communicating the printed and now the electronic word. The social component could just as well be called the political or ethical dimension in scholarly communication. The diminution of the scope of scholarly communication, its reduction in relative size to the publishing world as a whole, creates a dynamic within this community. The sense of importance is not reduced because the market share is less than it once was. Quite the reverse, the sense of obligation to the higher purposes of this sort of activity becomes all the more relevant.

The purpose of publication, in its most elementary form, is to establish a communication between author and audience. We should remind ourselves of this simple purpose with respect to the moment in the twentieth century when scholarly publishing first encountered the new technologies when they first became introduced into academic life. This development changed the relationship of power between author and publisher, but it did not change the ultimate purposes of serving the larger community. My concerns in the late 1970s were focused on the disappearing "mid list" of commercial publishing, and the added weight assigned to scholarly publishing in circumstances of a less favorable marketplace. Over the past period it is evident that the competitive tendencies in the marketplace have intensified, with not only fewer books of scholarly merit being produced by commercial houses, but university presses attempting new ways to survive and thrive in a highly competitive, seemingly static, no-growth marketplace. How our particular part of the book publishing forest has sought to cope—whether through regional studies, the publication of fiction and cookbooks, or specialized emphasizes on marginal, even quirky, areas of social life that rarely find their way into broad markets— formed the private background for my 1979 address before the AAP (Association of American Publishers).

If we look at the situation of the 1970s as a timeline, I would have to say that the most important factor is that the world of scholarly publication has shifted dramatically from a commodity basis, the manufacturing, sale, and disbursement of physical entities called books that embody ephemeral matters called ideas, to a communications systems, a world in which proprietary considerations have profoundly shifted from the physical to the electronic, from the book as such to the information basis they have in common with all other media involved in communicating ideas. It was not that long ago that the number of copies produced defined the expectations of a particular title, and such arcane notions as advances and royalties. That clearly is no longer the case. Small print runs in the hundreds and rapid reprinting have seen to this change.

Having acknowledged what is now common wisdom—the shift in fault lines of scholarly publishing from the book or journal as the object or end unto itself, to the information contents therein contained—it is also evident that the social and ethical grounds of scholarly publishing remain very much intact. The tripartite system discussed remains firmly in place, and will remain so as long as creative intelligence remains a driving force in human relations. Scholarly publishing has never been entirely defined by either the physical property it produces or the profits it provides, but by the utilization of the ideas contained in those properties. The same exists today, it is not the existence of blogs and bloggers, electronic devices to enhance or engage the viewer, but the spread and utilization of ideas.

Such dramatic advances in the technology of printing have tended to exaggerate the technical and diminish the intellectual in studying the situation in scholarly communication. What has been defined as "special effects" has at times overwhelmed the publishing world. Everything from types of packaging, layout, design, three-dimensional image attachments, test examination programs, to the very format as such, have been introduced to increase the size of the market, that is, the number of sales. Curiously, in this process, the scholarly presses themselves have become lonesome repositories for the old-fashioned way of increasing sales: better intellectual products.

The scholarly book and the professional journal—whether in mechanical or electronic format—is the repository of ideas. It is also the trigger for new ways to explore that repository in the policy arena, and change them to meet the needs of living people. The new developments have not changed the elementary facts of scholarly publishing, but it has enhanced what we do and how we respond to that ubiquitous invention called the "end user." We the publishing community may provide, they are the people that still decide. That is called the democratic way!

The first, and perhaps fundamental question is one of comparisons: what, in fact, is the university press like and unlike? More specifically, how is it distinguished from commercial presses? There is a fair enough consensus on these questions to move beyond the chicken and the egg of this. Clearly, university presses have never been entirely distinguishable from commercial houses in that the latter occasionally publish serious books on subjects of scholarly interest. Sometimes both publish the same authors. But today, university presses publish some of the same kinds of books commercial presses publish for the trade—cookbooks, poetry, novels; books having to do with all kinds of esoterica. Certainly we can no longer simply distinguish university presses by the character of the books they publish. Indeed, to walk around any display of their books is to be assured of that fact. Many university press titles could have been published by commercial houses; indeed, some presses enter into joint publication with commercial firms.

We come down to the fact that university presses are different by nature of their organizational structure. They are part of a university, accountable to people within an environment in which making a living is important but not the only important goal. Furthermore, while they increasingly have fiscal responsibilities within the university, the board to which they are responsible is not necessarily comprised of stockholders, but rather a board of scholars for whom financial statements may either be anathema or unknown. At some level, university presses are responsible, as is any other part of the university structure, to a world remote from finance or financial concerns.

Such non-economic factors account for a large part of the challenge of university press life. Making money may be one key objective and it is certainly important for many smaller university presses. But in university press life, troubles brew when bad books are published, or when books are reviewed negatively. Such troubles may occur in various guises: removal or reduction of subsidies is the key outcome. Trouble results, on the one hand, from failure to meet sales objectives; on the other hand, it results from not producing books that are qualitatively superior, or assessed as qualitatively superior to their commercial rivals by an internal university board or by the academic community as a whole.

In many ways, university presses disguise problems of size, or bury them within university contexts. It is an anomaly, but one can have a large university and a relatively impotent or small university press yet a relatively small university with a powerful university press. There is no automatic correlation between the size of the university and the size of the university press, or its power. Indeed, some rather outstanding and highly renowned universities do not even have a university press. Some renowned universities have had university presses go under much as they do in commercial life; they fail as businesses, so automatic correlations between university and university presses do not operate. We are dealing here with an uneven distribution of policy and talent no less than profit and service.

This observation is in many respects commonplace and known throughout the publishing community. Let us then see what it is that gives special character to scholarly publishing by university presses. In part, I would have to divide the world of scholarly publishing into feudal, capital, and social sectors. These operate simultaneously within university presses and help create the character, the tension, and the strain within university life. Let me outline this tripartite nature so as to make clear the point.

First, there is what might best be termed the feudal or the medieval element: the prestigious or the status element in publishing. The feudal element books are physically attractive, often very expensive to print and to bind, and they represent a best foot forward in presenting an image of the university and the university press. These kinds of books are produced, very often, in full knowledge that they may never recoup the investment. These prestigious books, these status books, must be published in order for a university press to exist, not in commercial terms, but as an ideological entity in the world. This medieval element plays an equally strong role in the university itself.

However, a second layer of truth comes into play in regard to the press's position within the university: the university press is also involved in the area of capital. It is involved in the area of profit, profit-making, and the marketing of books which, however feudal or medieval in character, must be sold with advanced entrepreneurial concepts and methods and a keen awareness and competitive appreciation that there are other publishers and other books. To demonstrate this, I cite a University of California Press advertisement which appeared in the May 1979 issue of *Transaction/SOCIETY*. The copy reads: "California brings you overcrowding, lesbian communities, concentration camps, mental illness, and other pressing issues of the day." In each case, these cutlines refer to books that happen to be outstanding. The three I have read, by Mark Baldassare on residential crowding; Nancy Scheper-Hughes on mental illness in Ireland; and above all, the extraordinary analysis of community devastation and restoration in Auschwitz by Anna Pawelczynska, are entirely commendable. But each could have been marketed in an infinite variety of more sober but possibly less memorable (hence less effective) ways.

The nature of their books, the medieval character of the university press, the search for prestige on the part of that press, does not prevent the capitalist marketing of its books. How the press markets, how it presents that book has to be done in consideration of the marketplace, not only of ideas, but the marketplace in its strictest commercial sense. This second layer of truth is what calls into play very intense feelings about the kinds of books that are being published these days. Such feelings bear on the medieval character of the great university presses and their books. Because as financial considerations penetrate, permeate, and percolate through the university system, then the issue of whether university presses should publish cookbooks, self-help books, and self-actualization books, becomes very important. We are not simply talking about book one. Any university press can rationalize and justify a cookbook from, let us say, the People's Republic of China. But the twentieth cookbook on the list becomes significant, since it has undoubtedly pushed out some of these other books, which I call medieval, which you may call prestigious, or which others may call status books.

Therefore, questions of the character of a publishing list, and the nature of entrepreneurialism, are not incidental to university publishing. Nor are they the only questions involved. Negotiations between the feudal and the capital sector can help one define the character of a particular university press list. But even this second layer of truth is rather in the realm of the commonplace, rather than what most publishers know, understand, and at least operationalize, even if they do not understand the ephemeral character of the bottom line.

The third element is more complex. It has to do with the social psychology of people in the business of publishing, and that includes myself as well as many in this audience. This third element is what we might call the egalitarian, social equity, or one might even say, socialist or anti-capitalist bias of many people who enter into the field of publishing. One of the original attractions to publishing for many people is a desire to get away from capitalism, to get away from the feeling of the rat race of the ordinary work-a-day, to get beyond entrepreneurialism in its raw, naked, competitive form. That socialist impulse and anti-capitalist impulse, is a very important element in the character of university press personnel, as well as in the character of the organizational structure of university press publishing.

University presses have many employees who share the values of the university, but they also share the values of the vanguard of the population as a whole in terms of women's rights, the rights of black people, minority issues, deviant behavior, and what have you. So the social component, when added to the medieval and capital components, really provides three levels of interpretation that express the character of university press publishing. It is a mistake to ignore the third element, or to ignore the social psychology of the publisher and the directors and the association directors, of every one involved in this area, not only because of the special ideology of the people involved, but because at some point the people who enter the world of publishing to escape considerations of the market, are faced with making decisions about that market. How they handle those choices also defines the nature of a press.

This anomaly between altruism of self and egoism of commerce is keenly felt. It is internalized and anguished over, and cannot simply be ruled out as a factor in how publishers work. For example, most university presses are recognized as having certain types of lists that are particularly important. The University Press of Oklahoma has a strong bias in favor of Indian people and Indian culture. Oxford University Press, because of the editorial interests of its directors, has a strong impulse toward American history and toward larger problems of world civilization. The University of Chicago Press is bound up with the history of problems in normative political theory, social work, and sociology; it reflects the special, rich history of the university and the city. Yale University Press has been very strong in areas ranging from urban analysis to military history. Because of its linkages with the Bollingen Foundation, Princeton University Press has an outstanding list in psychoanalysis, art history, and religion. Smaller university presses like Howard emphasize third world studies. One can multiply these special interests by the number of AAUP (American Association of University Professors) members. These characteristics do not simply reflect the random submission of manuscripts. Specialization derandomizes the selection process within the publishing houses themselves. The directors, the personnel who have strong commitments and strong beliefs over and above entrepreneurial and even status considerations, also give a press its character. In other words, a press may end up publishing a low status book by a low status author without hope of entrepreneurial recoupment of investment because it meets a social need that the press has felt for a long time and with which it is strongly identified.

The coexistence of these feudal, capitalist, and social elements helps explain what the university press is. Having so described it, what then is the bottom line for the university press? Now clearly, even though it might be an exaggeration and oversimplification, in large-scale publishing the first and the second questions are, did you make your numbers? Did you make the profits? And did you meet the profit expectations? Did you raise the standards of profitability for the year ahead, or for the next projection period, whatever that might be? But in the university press environment, the mix between the three factors just described determines whether a firm is adjudged successful or unsuccessful in meeting its goals. So that the mix is the essence of the difference between the large-scale commercial publisher and the university press and the small, independent scholarly and academic publishers that we identify with the university press.

Non-economic factors that are adjudged indicators of success are difficult to quantify, but with the rise of a social indicators movement, hardly impossible. They become extremely important to the directors and the associates involved in university press publishing. For example, did a particular book result in legislation is a question that can be asked. Did the book get picked up by the mass media and receive special attention in *Time, Newsweek,* or *Fortune*? Does a book form the basis for widespread further discussion at subsequent professional meetings? For example, the debate on E. O. Wilson's sociobiology text which Harvard not only published, but vigorously marketed, illustrates the linkage of a book to the inner history and paradigmatic character of the social sciences.

Does the author of a book become a figure of note in his or her own right, an expert who endows a particular publishing list with a dynamism based upon esteem rather than just profit. Whether or not the book sold well or did not sell well becomes of secondary significance. Again, these non-economic factors can be quantified, and can be measured. They should certainly be included in any calculation that a university press makes about the success of its list, because the university press is addressing itself to issues that have a great deal to do with the future of the country and the future of the world.

Having said this, what constitutes the strength of the university press and the scholarly press community? Here I would say it is exactly its plurality of interests and themes. It is also in the plurality of university press ideologies. One again must be frank, especially in the age of détente and reconciliation with the socialist bloc. We are not a national community like China or the Soviet Union. A political system does not determine our publishing programs. As a result, it is neither possible, nor feasible, for our press directors to speak to corresponding personnel in the communist bloc with the same degree of assuredness or implications of a national consensus. Precisely because the U.S. does not have an ideological unanimity, the university press becomes an important phenomena in another non-economic aspect of life: namely, the problem of democracy itself. The publishers of university presses are involved in the question of democracy, not simply in the question of the characteristics of their publications.

It is perfectly fair to say that one must make a realistic appraisal of economic factors in order to survive. Anyone at Northwestern University Press or Case Western Reserve Press could have told you that. On the other hand, we have to remain aware of the dangers of the kind of determinism that results from an over rationalization of economic factors. There are things that one can do as a university press that cannot be done in China or in the Soviet Union, and the truth is that this is good. The moral purpose and the moral value of the university press is outside the purview of any single national plan or policy. The university press is one foundation point of cultural democracy in this country. It is dangerous to trade this away for pure economic determinism or a pure economic deterministic theory. Therefore the ultimate value of the trilogy of medieval, entrepreneurial and social factors, what makes them work, is their coexistence. The coexistence of the seventy-five to a hundred scholarly presses in the United States represents democratic values in their very essence, in their very existence, through their very publishing programs. The university press network transcends in importance any presumed organizational unity. That is important only in the narrowest sense. It is in their diversity that university presses are important, for in their diversity they represent the democratic horizon.

4

Limits of Standardization in Scholarship

The publishers of academic journals now must pay more for printing and binding at the same time as they are reducing print runs. They must also farm out the of purchase expensive equipment to engage in small-run printing. Sales are dropping for a host of reasons. Some of the reasons are academic—ranging from new fields of specialization to declining enrollment in some older fields of enquiry. Others are technological: new modes of reproduction have reduced the number of individual subscriptions and increased the use of institutional, library, and inter-library services. There are also non-market factors, such as an increase in the number of journals to meet the severe competition for tenure as well as the reporting of research breakthroughs. The strain placed on publishers of scholarly journals has been considerable. They have taken sometimes draconian measures to cope with these pressures and reduce, or at least keep constant, their unit costs.

One area that has received increasing attention is standardization—ranging from copy editing through printing and binding. The nature of the new technology, specifically of word processing, makes it eminently feasible to arrange for systems of production based less on journals or books per se than on articles or chapters as the unit of measure. With sound office management, articles can be typed into the computer, copy edited, and revised in ways which minimize traditional delays in waiting for the last article in a journal or for a recalcitrant editor to return materials to the journal offices. The concentration of journals in the hands of relatively few publishers is more a function of the need for standardization than of any innate impulse towards monopolization. It has occurred simply because in every phase of publication—from manuscript processing to subscription fulfillment—it has become apparent that economies of scale require some sort of concentration. Concentration guarantees efficiency of production as well as of post-publication fulfillment.

In whatever direction one looks, standardization seems to be the order of the day. The age of a single university or department within a university, or even a professional association engaging in all phases of production, design, and delivery of scholarly journals has drawn rapidly to a close. Scholarly presses have assumed a major part in this endeavor—and in so doing subject the journals to a division of editorial and production labor formerly reserved for books.

But, as is often the case, the *necessities* of the commercial and academic marketplaces (and they are not always synonymous or synchronous) are converted into the *virtues* of the publishing process. The argument is usually made, and not without persuasion, that the key to scholarly journal publishing is informational content. The consumer requires data, formulae, theories, rapidly and accurately. Further, given the small number of buyers and users, a choice must be made between high standardization and still higher prices. Given the near-universal resistance to high prices, standardization is usually seen as the best, at times the only, way of delivering information to a stratified academic marketplace.

I would not deny the world of standardization its due, but I do assert an equal need for diversification, or if that proposition be too strong, I would argue the economic as well as aesthetic values of diversification. Standardization in fulfillment services may be an absolute necessity, but standardization of the size and shape of journals may prove to be a much lesser blessing. What follows is not an argument against the present in the name of a gentlemanly past, but a commonsensical recognition that the appearance of diversity is still very much part of the reality of the publication of scholarly journals and books.

The issue of standardization vs. diversification exemplifies the fact that what initially appears in publishing as prosaic may, in fact, raise profound considerations. Everyday decisions about costs of production and plant overhead embody issues of far greater magnitude that must be faced by every publisher, and by none with greater urgency than scholarly publishers. Some might protest the open-and-shut nature of this topic in the name of the bottom line; but self-imposed, or at least non-market, limits have already moved academic publishing into a terrain far removed from pure economic strategies and commercial policies. The alternative is to risk the issuance of hard copy publications in favor of purely electronic delivery systems. In the absence of diverse print and format approaches, such a dicey outcome is not only possible, but even likely.

The search for a balanced approach between standardization and diversification takes a different form in different segments of academic publishing. Journals, books, and encyclopedias each have their own limitations with respect to the principle of standardization. Obviously, all publishers search for standards in a pure sense. But diversification is more than a costly necessity. There are both practical and metaphysical reasons for diversification. Academic journals are not simply economic goods. Each scholarly publisher must factor in diversification as well as standardization—and do so in an articulate and viable way.

The first thing to be observed about journals is that, unlike encyclopedias or books, they are not an entirely unitary phenomenon. They differ profoundly in character, frequency, nature, and purpose. Journals are sequential and consequential while reference works are individual and episodic in character. Let me therefore move to a consideration of ten other areas where academic publishers need to strike a balance, or at least seek to strike a balance, between uniformity and diversity, standardization and differentiation.

The first of these is a function of paid circulation. One cannot have a journal with 250 subscribers, predicated as a translation service from a foreign language directed to a highly specialized audience, come out looking like *Fortune* magazine. For journals of such minute size but of service to a vitally interested community, the publisher need not even worry about adjusted margins. All that is needed is a translation in a format that is relatively sensible and intelligible. Such a journal is extremely different, by the very nature of its circulation, from *Fortune* or *Forbes* which reach professional audiences in the hundreds of thousands. Clearly, one limit to perfect standardization or uniformity is circulation.

A second element is allied and related: namely, the price asked. Again, in part, this is determined by the subscription base. If one publication has 250,000 subscribers and another has 250, the price that a publisher can ask is as different as the expectation of the ultimate consumer. Both publishers and readers expect certain standards of design and format based on the cost of a subscription. In other words, the purpose for which the journal is being bought is a powerful determinant of the character or uniformity of that publication. In parallel with price, the number and nature of advertisements may affect a periodical's character. Certain journals—for example, the one mentioned previously of translations may carry no advertising, or at most only in-house advertisements. Such publications will obviously look and feel quite different from those with many display advertisements.

Frequency is a third factor. If something is published as a weekly or a newsletter it will look different from a quarterly. It will be on a different paper stock and it will have different unit costs. If a scholarly work comes out once a year, the publisher may release it in cloth. If it comes out four times a year, it may be Smyth-sewn but paperbound. If it comes out every week, it may be on newsprint, folded and gathered, or at best saddle stitched. All journals cannot be bound in the same way or assembled in the same way. Every publisher who wants to branch out into new areas—newsletters at one end, or annual serials at another—must come upon this problem. Diversity follows variations in frequency as a result of economics no less than of academic intransigence.

Another determinant of uniformity or lack thereof, is special content: charts, graphs, tables, and other materials that have special requirements for reproduction and thus affect the cost per page, as well as the appearance, of a particular journal. These materials cannot be legislated out of existence by publishers' urgings. Certain kinds of charts, graphs, and tabular materials will look better on an 8 1/2 x 11-inch page than on one 6 x 9 inches. The size of a journal is a response to the physical representation of data that a publisher must include for readers of a particular profession or a particular organization. Thus, non-linear material is a significant limit to any notion of uniformity.

A fifth element involved, and one easily overlooked, is the role of professional organizations. Their goals may not be the same as those of the publishers. Many professional societies have their own notions of what they consider appropriate. The professional "bottom line" has its own dynamic. Whether there is an isomorphism between a society's requirements, or perceived requirements, and those of its publisher must be determined in the play of forces at work. It is no accident that certain kinds of journals attract certain kinds of professional respondents by virtue of appearance. The general appearance of journals varies with professional codes. Such professional requirements limit any strict uniformity that publishers may wish to impose.

A sixth element is what might be called a differential manual of style. How is a scholarly publisher to react when confronted, while in pursuit of uniformity, with the anthropological, psychological, and sociological manuals of style? Or, beyond these, with the styles of the physical and mathematical sciences, each of which may differ considerably from the *Chicago Manual of Style*? If a publisher insists on telling anthropologists that their journal must look like a psychological journal that publisher runs a risk. Ultimately, the specialists or their organization will think less of the final published product or will remove the journal from what is perceived as a recalcitrant environment. This is not simply a matter of taste, but one of organizational notions of standards that have been developed over many years. Publishers have to deal with multiple concepts of standards, not simply publishing standards.

The seventh factor militating against standardization, and perhaps the most difficult, is aesthetic. Every journal involves design: type, margins, crossheads, title and author identifications, extracts, and so on. A journal looks like what it wants to be. The aesthetic dimension represents antecedent histories in a given area of research. If it is a new journal of foreign policy, it will want to look like the other five leading journals in that area. In its search for scarce resources—good manuscripts—it will not want to look like the *Journal of Abnormal Psychology*. There are good reasons for this. The scholars of international relations do not want it to be implied that foreign policy is abnormal (although what it is in fact is another matter). The aesthetic dimension is not to be entirely discounted, nor is it entirely aesthetic. What constitutes evidence, and the way it is reported, is of the scholarly essence, although it may appear as idiosyncratic and artistic in the obvious sense.

From this it follows that a publisher must defer to the scholar in determining whether a journal may appear too simple for its purposes, or too slick for its purposes, or just right. The scholar does not want to have a journal like *Foreign Affairs* looking like a newspaper. The more volatile the subject matter, the more certain journals will insist upon an appearance which is tranquil, cautious, even reassuring. The publisher must work with the profession to ensure a psychological balance; a publication which a consensus of readers believes captures the right image for the profession. Discipline journals sometimes fail not only because of weak contents but because they are too slick or their covers are too bold or wrongly designed for their purposes. Getting physical appearance and intellectual purpose together is a publishing function, one that crosses over between economic and editorial considerations.

A further element, one we have been hinting at, is the journal's own history. Very often publishers do not start journals but inherit them. There is a presumption in many discussions about standardization and uniformity that publishers begin journals *de novo*. But more often than not, they take over journals from other publishers. The journals have their own histories; and with those histories come a baggage of peculiarities in performance. This does not mean that everything that journals once did in the past is correct or sacrosanct; neither is everything they have done in the past incorrect or devoid of value. Publishers must be sensitive in their dealings with the vagaries of a journal. Historical antecedents are an important limit to uniformity. To ride roughshod over that inheritance, that collective memory is to ensure contention between journal editors and journal publishers from the start.

The penultimate point I would make is that publishers of academic journals, at least in part, are dealing with the priorities of marketing; they are selling a precise periodical to a precise external audience. They are doing everything possible to emphasize the journal, not the publisher. This emphasis on periodical rather than publisher is expensive but offers a benefit: subscribers are never under the false impression that if they read one journal, they know what is in other journals published by the company.

The final point is that uniformity is a function of specific contractual arrangements. Publishers do not simply have carte blanche. A journal, its editors, and its boards enter into a fiduciary relationship with a publisher. Responsibility between the publisher and the journal is a negotiated order of things, settled upon in the crucible of supply and demand. In that contract are written certain parameters beyond which the publisher really cannot go, regardless of any arrangements in effect with other journals in the same publishing program. Contracts too are a limit to uniformity. The contract is neither a defense nor a critique of uniformity. A contract represents the collection of interests necessary to realize a journal as an end result. It defines limits to editorial differentiation, but also to publisher integration.

There is indeed a bottom line, a profit-margin which is in large part determined by levels of standardization. But the top of the line—awards, emoluments and what publishers can charge for its products—has to do with diversity. Publishers are rewarded at two levels: at the level of economy and at the level of status. It is good sociological sense to realize that both elements are needed for profitability The system of marketing is a functional response to the uniformity of an area, but also to the distinction which each periodical uniquely and individually represents in that field of professional endeavor.

In the world of journal publishing, the issue of standardization and diversification is less either/or than both/and. It is in the tension, the dialectic, between uniformity and diversity that publishers and editors alike must fashion a meaningful product. And let it be noted that I use the word meaningful. For even the most hard-boiled publisher, intent on maximizing uniformity at all costs comes upon the ineluctable truth that scholarly publishing is itself a limiting factor to profitability. Yet I have never heard of a publisher of scholarly materials opting for pornographic publishing simply to honor an abstract notion of bottom lines and profits. In this curious commitment to scholarship and professional quality lies an essential solidarity that reduces the polarized decision between standardization and diversification to a modest, everyday sort of problem.

With all due respect for Weberian notions of rationalization, the publishing community must take into consideration that the very nature of the book or the journal includes aesthetic dimensions: everything from the size and variety of typography to design elements that are embedded in the nature of the product being produced and offered for sale. To neglect such "non-rational" elements, to reduce everything to the content of the work or the message is to hasten the end of regard for the printed form of communication (whether delivered in mechanical or electronic form). Indeed, it should be noted that from its very origins as a painstaking product of handiwork, the book and the journal have included a variety of design presentations that have a value added component. So before the publishing community entirely turns over its tasks to information technologists lacking in aesthetic vision, it would do well to keep in mind the fate of all purely mechanical objects lacking in a sense of the transcendent, and at times, of the metaphysical no less than the physical. Even the useful artifact, such as automobiles, ships and airplanes, must be graced by the beautiful. That is the final equation in the process of rationalization and standardization of ideas.

5

Publishing, Property, and Information Structures

Upon reviewing the *Preliminary Draft of the Report of the Working Group on Intellectual Property Rights,* one immediately confronts the grand ambiguity that resides in the two words: "intellectual property." That the National Information Infrastructure had to locate precedent for its missioning Supreme Court Justice Story's 1841 observations on copyright issues as an area involving the "metaphysics of the law" indicates what a long reach the very notion of intellectual property entails in a democratic society.

Metaphysical considerations notwithstanding, the legal profession has managed to operationalize the field of intellectual property with amazing acuity. This is now a mushrooming activity that covers trademarks, trade secrets, unfair competition, copyright and copyright infringement, trademark prosecution, licensing infractions, and property audits. In addition, an entire area of law now covers computer and high technology as such, including areas of hardware and software integration, copyright, patent and trade secret protection, and broader issues of computer security and privacy. This is only a portion of the activity included under the rubric intellectual property. What follows is not an attempt to deprive the legal profession of new areas of activity, but rather to raise the possibility that the area of intellectual property is in fact old wine in new bottles.

This statement is not intended to be a broad survey of the legal and technical issues covered; I will make one simple point:

the concept of intellectual property is inherently fallacious, and its consequences are dangerous. I do not deny the obvious: (a) a national information infrastructure already exists; (b) integrative mechanisms are at work at technological levels that will eclipse present delivery channels and systems by a considerable margin; and (c) that such developments have substantial potential to increase public access to information and entertainment. The challenge to those who create and disseminate such information is how to prevent unauthorized uses of their works. A wide-ranging series of legal suits, some still in litigation, have made it clear that technology brings with it ambiguities and controversies. This is clearly outlined in the Green Paper. But the fact that we have had a continuing round of hearings spread out over years makes it clear that time and technology by themselves have not been able to dissolve older concerns, nor has the law been able to adjudicate and resolve standing issues.

That the National Information Infrastructure dramatically changes the way information is prepared and disseminated is no longer at issue. Whether we are dealing with online services or new forms of electronic publications, the dissemination of information beyond traditional print modes is clearly with us. More troublesome in the report is the phrase that "the degree of distribution of or public accessibility to electronic documents is not presently measured and may prove immeasurable." The last clause causes great anxiety: for if we are dealing with a phenomenon inherently immeasurable, then indeed, it will take a great deal to restore a sense of calm to the variety of information and communication industries impacted by such new developments.

By extension, such logic of ambiguity could apply to holdings at every municipal or state library, which also depend on the taxpayer base for survival and purchase. Given a viewpoint of the non-measurable nature of intellectual property, not a single book or journal could expect to retain present copyright protections—much less extend such protection as publishers and authors are now seeking. By extension, if no protection exists, then the long-run consequences are a drying up of the creative processes that go into books and journals, or worse, their publication as private documents that must be purchased by individuals, and that will become thoroughly unavailable to patrons of libraries and allied institutions. This is a disastrous dead end, not an information highway.

If we are to break the logjam that presently exists, or better the dialogue of the deaf between the legal system and the technological infrastructure, we need to move beyond the confines of both. This does not require that we move backwards toward metaphysics. It does mean we must move toward linguistic clarity and if need be, ethical judgment. For discussions on controlling access at the server level or at the file level, or for that matter controlling the use of the work itself, all strongly imply—quite correctly—that the technology in place today far outstrips any capacity to prevent illegalities from taking place. Indeed, even when confusions and ambiguities in the law have been laid to rest—as they have been in a series of legal rulings affecting copyright—violations by good people still occur. These are ordinarily honest people who would be shocked and appalled to learn that they are stealing the property. I submit that such theft will persist, whatever legal decisions, and future hearings guaranteed, as long as the concept of "intellectual property" itself remains unchallenged. For this phrase is the soft core, the loophole of loopholes, that permits ordinarily good and honest citizens to engage in illegal actions, reproducing everything from software programs to musical scores without authorization or compensation to the copyright holder.

The critical problem resides in the very notion of "intellectual property." The addition of the word "intellectual" is mischievous. Property is property, and this is what is germane to the analysis of how copyright and patent property rights are to be protected in this new informational environment. Indeed, the word "intellectual" permits otherwise honorable people to dismiss their participation in a normative framework supporting illegal usage of copyright materials.

Publishers and authors disseminate and need to protect real property. Books, journals, serials and magazines are products sent to market. CD-ROM on disks are likewise products sent to market. It is time to break the cycle of clumsy distinctions between physical and intellectual product as if the former were worthy of protection, that is, as hardware; while the latter are in a different and lesser domain. It is also time to break the myth that librarians seek dissemination of "intellectual" property while publishers are only concerned about protecting it. Both communities face common problems of survival; and as a consequence, have shared requirements for growth.

The need of the hour is to move into an economic mode of analysis. Publishers and authors need to be compensated for their efforts—just as plumbers and waitresses need to be paid wages for their efforts. Librarians must find a way to secure revenues from customers if they are not to be constantly thrown back on the notion that they must uniquely look to the taxpayers as a source of funding. As things now stand, inadequate funding means that libraries are abandoning their responsibilities to archive while investing heavily in technological devices which may provide a solution to their problems of space, but may or may not address problems of service to their customers or problems of inadequate funding.

What is ominous is the huge expenditure of capital by libraries for new technological devices that offer abilities to network, reproduce, and retrieve without even the slightest recognition that property rights are being grotesquely and manifestly violated. Libraries appear to be proceeding under the assumption that "intellectual property" is somehow unique, and hence not subject to the normal rules and rights of the marketplace. For this reason, we must abjure pleasant conversations about intellectual property, and return to basic core concepts of property. Our focus must be how we can increase dissemination without destroying protection.

The distinction between physical and intellectual property dates back to the ancient Greek notion of a separation between manual and intellectual labors, between what used to be known as head and hand. This distinction was serviceable as long as firm distinctions between those who serve and those who are served were enshrined in the common culture. Even the great Aristotle took for granted that creativity could take place only in an environment in which manual labor could provide essentials for all and released time for the few. We no longer live in such a world. The new technology, the national information infrastructure is itself evidence of how fluid arrangements among people have become. We live in a world in which engineers, technicians, psychologists, and hosts of maintenance personnel speak with each other, and learn how much they have in common, precisely because of the pace and scope of science and technology. Yet we still employ a rhetoric of intellectual property better suited to ancient Athens than modern Washington.

Once arbitrary and fictive distinctions between types of property are eliminated in our thinking, the issues shift as well: what sort of property protections do publishers lack that are available to other enterprises? I think the answer will be few. That is to say, the law is quite specific when it comes to issues of theft of goods or denial of attribution, or for that matter forgery and plagiarism. The need of the moment is a new culture more than new legislative relief. Technology may create possibilities for infringement that require new legal interpretations. This is a far cry from the need for yet more legislation. The national information infrastructure is a culture. It emerges, as the Green Paper itself makes abundantly clear, without much regard to the status of the law. It is not a mandate from heaven so much as a phrase that seeks to capture that which is unique in the world of information and communication as we enter a new millennium.

We must deal with the issues in which technology is outstripping legal protection and how proprietary claims as such are under assault. Few would argue against an information superhighway or its legal rationalization. What is being determined as we speak is how rewards for the creative processes involved are to be allocated. The law itself is perfectly clear on this. As the Green Paper indicates, "the legislative history to the Copyright Acts makes evident that any form of dissemination in which a *material object* does not *change hands* (the emphasis is in the original) is not a publication no matter how many people are exposed to the work" (123). The converse is also the case: copyright involves proprietary claims over that which changes hands—and that is subject to laws that govern the exchange, rental or lease of real goods, real objects. The recommendations go a long way to enhance this distinction between property and performance. What are being transmitted for sale are recordings, books, papers, or copies of material that constitute publication. The public performance of an eighteenth-century chamber work by musicians is not at issue; the performance of an adagio for strings by the twentieth-century composer Samuel Barber is at issue; the American Society of Composers, Authors, and Publishers collects royalties on behalf of the artist. What is at issue in relation to Mr. Barber's work is the photocopying rather than the purchase of the score. This is also a factor for music publishers who produce new scores of the works of eighteenth century composers. The conclusions of the Working Paper go a considerable way in recognizing such distinctions. It reverses gears, and raises anew issues of legal and technological protection and their limits. This is wrongheaded.

In its discussion of fair use, the Green Paper proposes extending the range of fair use so that the poor and disadvantaged are not disenfranchised by the new information revolution. In point of fact, the problem is hardly the poor and disenfranchised, but rather the rich and greedy. For we are dealing essentially with white-collar crime in the theft of material from books, journals, papers, and the like. Public policy can make provisions for those who cannot pay for these materials just as policies ensure the poor receive food stamps. Indeed, I see no reason why a person entering a supermarket should not be permitted to use such stamps to purchase books on the shelves as well as bread on the shelves.

Here the question of the value of property becomes central. How much is a specific piece of property worth in *specific* contexts, and further, how much should it cost in relation to other forms on which it may be available. In the past, once copyright holders ventured outside conventional delivery mechanisms (print, film, and audio) their share of revenues was miniscule. Those who produced the hardware and those who engineered the delivery system reaped the lion's share of reward. This was possible as long as the income streams from these sources did not cut off major supplies of capital from sales of traditional products, and as long as the new information technologies did not directly impede the survival of traditional publishing as such.

We are now involved in a world in which the struggle for property spills over into a bitter struggle over who has the right to disseminate property—or beyond that—seek a reward for such dissemination. The new push by the Library of Congress to make millions of documents available to the wider public illustrates present strains: For if the library permits users to reproduce in a variety of electronic forms materials that are new as well as old, privately published as well as federally issued, and organized in ways that may or may not be approved of by publishers or authors, we are in dangerous waters. The rationale for such dissemination is that the taxpayer "buys" Library of Congress materials; hence the taxpayer has a right to see and use such material assembled at the Library of Congress. In fact, the library requires donation of all materials for copyright purposes; the publisher, not the taxpayer, pays and has no choice unless he wishes to forego copyright.

We are thus entering an environment in which old allies become new adversaries—and vice versa. Media Labs at the Massachusetts Institute of Technology, Electronic Data Systems, a division of General Motors, Xerox Corporation, Bellcore, and the National Science Foundation, are a diverse group indeed. They have in common an interest in both sides of the information highway: They supply hardware components to feed the highway and own valuable copyrighted materials that need protection. In this environment, a certain cannibalism is inevitable: large-scale publishers capable of making the investments required to drive on the new information highway will do whatever they can to increase their monopoly by incorporating materials generated by smaller publishers. For without access to that which they do not own, even the largest publishers will be in the position of television networks without material to air. Smaller publishers for their part will need to reconsider the basis of contract as well as cooperation with larger publishers. In short, we have here a situation not just pitting technological breakthrough groups against conventional publishing groups, but one in which the players are themselves confused as to their corporate identity and corporate requirements.

The government is likewise in this condition. It must advance the cause of the widest possible access of information to its citizens, while protecting the sources of creative energies with the society. Unless the dialectics of the moment are understood, such hearings will be followed only by more hearings, and the fundamental issues will be obscured. We need greater clarification of concerns, followed by policy guidelines that can be understood as universal in scope and fair in purpose.

Intellectual Property and Internet Publishing

Some time ago I received the following letter (whose origins shall be kept private in the name of intellectual property and the ghost in the computer machine):

> The University Graduate School is considering requiring theses and dissertations to be submitted electronically via the World Wide Web. You are the editor of a journal (Society) in which graduates from the Department of Economics may desire to publish. If a student's thesis or dissertation were available on the web, would this preclude the student from submitting an article based on this research to your journal?

The situation of this letter clearly needs clarification, especially in view of the fact that certain professional journals, especially in cutting-edge fields of research, appear to be taking the view that the release of preliminary versions of articles on the Internet does, in fact, preclude later publication. While being cognizant of the fact that different professional fields may offer different responses, my own view leans in the opposite direction: namely that release in cyberspace is not in itself a sufficient reason for declining to publish in hard copy form.

The student, the professor directing the dissertation, the department sanctioning the product, and the university which the student attends, are at liberty to put up on the Internet any article, essay, thesis, dissertation, or other form of work, presuming that, in so doing, there is no copyright claim that would preclude transfer to a journal or book publisher. I do not believe this is a decision which has anything whatever to do with the journal publisher or editorial board of a journal. The task of a scholarly publication is to separate the wheat from the chaff, or, to be more blunt, the garbage from the gold. Making a piece of work available on the Internet does not substantially alter that task.

A copy of most dissertations is sent to University Microfilms (UMI) for archival purposes. This does not constitute either an endorsement or a barrier to publication. I see no difference in qualitative terms between archiving a hard copy and making an electronic representative of a piece of work available. Indeed, in my opinion, the task of scholarly publishers will remain essentially the same in the new electronic age as it was in the old medieval scribe age: authenticating the quality of work being made public, whatever the format.

A journal is not a mechanical contrivance of typefaces. It is a human artifact of decision-makers, of gatekeepers, who certify quality, durability, and value. They cannot perform this task with special interests in self-promotion—whether they are students or the institution in which the work was performed. A limiting case, in my view, is a journal which is published on the World Wide Web, but retains the basic character of a refereed publication. In such instances, the only difference between a standard and a cyberspace journal is its availability in hard copy, that is, printed form. And that gets us to the heart of the problem raised by my concerned academic institution: that certain journals and professions have stated for the record their opposition to even considering work for publication that has been disseminated in hard copy. This represents a sorrowful confusion between form and content. It muddles the relationship between the mode of delivery of information and ideas and the substantive considerations that go into intellectual and academic cogitation. I fear that with the explosion of information in so-called cyberspace, this confusion will persist. But the history of publishing is also the history of self-promotion. So what we are witnessing is more a quantitative implosion rather than a scientific or cultural explosion.

The world of scholarship is comprised of a delicate set of relationships between sensitive souls: writers, researchers, publishers, editors, referees, and a myriad of individuals who function in a number of overlapping capacities. Before these differences between various good and decent people escalate and harden, I suggest that we look closely at differences between new forms of delivering information and traditional judgments as to the content and quality of such information. When we do so, our intellectual balance can be restored to the benefit of the entire community of scholars and publishers.

6

Specialization in the Electronic World

How does a small mission publisher handle such hot topics as electronic and/or on-line publishing? I suspected that our response to this question was minimal or casual, until I began to analyze it. I now share the approach that we have adopted in the hope that it may be useful to others.

Firstly, we acknowledge to ourselves the fact that electronic document delivery services constitute a whole new industry, not just a new aspect of our old industry. The parameters of this new industry, whether public or private, large or small, involve a different range of hardware, constantly evolving software, and a plethora of non-paper products, auditing techniques and monitoring services that are alien to, or at least distinct from, the historically evolved functions and structures of publishing.

Secondly, large corporations such as McGraw-Hill, who have pioneered the new electronic formats, are able to glide right into this new environment. Even they have had their troubles in figuring out what works, and what does not work, in commercial terms. But their very size gives them the infrastructure that permits wide latitude in experimentation and implementation. These competitive advantages of large firms must be recognized as durable and long-lived.

Thirdly, we recognize that the relationships of forces in publishing are more complex than they have been in the past. In addition to the new players in the electronic end of publishing, there are the old players, often publishers themselves, who have become involved in the new print technologies. Thus, McGraw-Hill's *Primis* (see LOGOS 2/4), a mechanism whereby end-users can go to a single publisher for their entire journal article needs, is a wholly owned subsidiary of the corporation. *Primis* guarantees a first usage payment fee and offers much lower fees to publishers for subsequent usages.

At the other end of the spectrum of bigness is Bell & Howell's University Microfilm International (UMI), whose former president, Joseph Fitzsimmons, made a point of "regarding publishers as our information partners." UMI has worked with notable success to uphold copyright permission agreements, publish copyright acknowledgements, provide a billing and tracking service for royalty payments and usage data, offer direct license agreements to their users and suppliers, and incorporate copyright clearance principles into their own educational program.

These examples of services offered by large corporations which afford small publishers return for the reuse of their materials without additional investment raise two questions. Will they be underwriting or undermining the work of the Copyright Clearance Center (CCC), which, after all, was a creature devised by the publishing community to address the problem of supplying the marketplace of ideas while protecting publishing interests? Secondly, how real is the notion of partnership? Will the participating publishers remain full partners of the UMI activity, once they have joined in, or will they become "junior"?

Until now, publishers participating in such schemes have accepted royalty formulas that amount to little more than a pittance—anywhere from 5 percent to 15 percent being the norm. In a full partnership agreement, publishers are likely to begin to demand fifty-fifty deals, at which point the profit margins of firms like Bell & Howell and Reed Elsevier, with whom we, along with other professional houses, have special working relationships, will be seriously impacted.

Smaller publishers, such as Transaction, cannot possibly compete with the wide range of services available in non-paper document delivery services—whether on disk or film or by telephonic devices. They are certainly not able to create a new environment for electronic use, which is the goal of McGraw-Hill, Reed Elsevier, UMI, and others. Having acknowledged this, small publishers then have to face the question of why they should have to enter a game dominated by the big fellows under rules over which they have no control—rules, indeed, hardly defined in law so far.

In solving this dilemma, it is important to recognize that the key to modern publishing is not the product, but the property. That is to say, the smaller publisher, when signing a contract for book or journal publication, includes all so-called subsidiary rights. Transaction Publishers have such rights because they run the risks—of the costs of marketing no less than printing, binding, and editorial costs.

The key to publishers' survival is controlling the proprietary rights. Once this is secured, on behalf of the author no less than the publisher, then the issue of how to parcel out such proprietary information can be examined. So the essential issue to start with is not how "Portia Faces Life" or how "Transaction Stares down Electronics," but what it is that a publisher actually controls in terms of contract, marketing, and production. Specifically, what sort of agreements can be negotiated both to the advantage of major publishers in the technology field and smaller publishers whose rights are to be renegotiated if they are to stay competitive?

Because publishers are small, this is hardly a reason not to be involved in the big issues. Transaction has, over the past decades, been involved in the reformation of U.S. copyright legislation in the 1970s; the establishment of the Copyright Clearance Center; and as one of the plaintiffs in the Texaco suit. In the current environment, publishers are able to operate on two fronts: the legal and the technological. Most attention has gone to the legal end. But small publishers should also support the technological efforts of such major firms as Xerox, who is developing encoding systems at one end and image enhancing systems based on intelligent digital copies at the other. I believe that such ancillary roles for specialized publishers (of whom there are many) will prove more beneficial than self-propelled efforts to enter the nether world of electronic communications (of which very few have become success stories).

So what is at stake is not simply how to handle the electronic environment or participate in the so-called superhighway of information, but how to negotiate these so-called subsidiary rights. As a result, the real revolution for a firm like Transaction is not so much technological as legal. We no longer divide rights into subsidiary and primary. There are only rights. And rights are a negotiable commodity that a publisher possesses at the point of contract and which may be surrendered—either through ignorance or volition—to others. Good legal advisers are more critical in such an environment than a new generation of super-fast computers.

Those who have taken the lead against misuse of their rights are often publishers of newsletters rather than journals, because their works are so easily and readily uploaded for computer distribution. The storing and sharing of newsletter data electronically have become widespread. One newsletter publisher, Fillips Publishing International, has sued a distributor, Atlas Telecom, for copyright infringement.

Document delivery in a plethora of ways has to be seen as a process that cannot be wished out of existence, nor is it one that we can hope will trip and fall. Rather we should wish more power to customized current course packets; more success to old fashioned microfilm devices and digital storage systems; more speed to the improvement of printing services through advance processing of electronic data. These services will be performed by non-publishers, as they have been in the recent past. However, this does not mean that the need for publishing will vanish. Publishing cannot be reduced to the provision of information. The old verities of judgment and intellect become more, not less, important. Document image processing may well be superior to print on paper in refining raw data, but it will not be capable of deciding which novel should be published and which rejected, or which scholarly monographs are truly innovative and which are patently imitative.

As a result, the heads of traditional publishing houses will find themselves closer to people in university life than to people in Silicon Valley. This does not mean that they will be able to avoid a keen knowledge of new technologies as they affect the cloistered world of information and ideas. It does mean that they will need to exercise sound judgment. Basic scientific and social scientific research in industry and in university life is not dependent only on daily changing data. The search for quality may even be hindered by the availability of quantity.

No small publisher can compete with the *Wall Street Journal, Forbes Magazine, Business Week* and their parent outfits such as Dow Jones on the delivery of daily information about the changing fortunes of the stock exchange. But these activities are no substitute for the production of basic books on economic theory and business cycles. No one suggests that Transaction should compete with Dow Jones. Yet, we seem to be pushed into the ludicrous situation of feeling that we have to compete with the likes of McGraw-Hill, John Wiley, UMI, and Macmillan at the level of documentary services.

What we should be doing, and what needs more attention, is the amount of fiscal return required to maintain essential publishing services. The existing formulas of 5 percent to 15 percent for parceling out rights to electronic forms of publishing are in need of careful overhaul. The loss of sales in hard copy form directly attributable to the new forms of document delivery must be taken into account when arriving at deals with CCC, UMI, Information Access, EBSCO, and others. No more are such monies "add-ons" to the stable business of selling hard copy. These ancillaries have become central to survival.

It is no simple task to determine the extent to which the new electronic forms of document delivery cut down the subscription bases of technical and scientific journals. No one disputes that there is an inverse correlation between the upward uses of electronic servicing and the downward numbers of hardcopy subscriptions. What no one knows is the actual relationship.

One indirect, and admittedly imperfect, index of subscription losses might be the patterns of personnel changes in libraries. Time Inc.'s library, one of the largest in the world, informed its staff in October 1993 that over a period of six months forty-three positions, one third of its library workers, would be eliminated. The stated reason was a shift from "paper-based operation" to an "electronic information system" made necessary by the needs of

Time researchers and writers for better cross references. One can hardly escape the conclusion that this development in personnel was quickly matched by a reduction in hard copy products.

It is not impossible that at some distant point scholarly publishers will have nothing to sell but their scholarship. Hard copy sales of a journal or annual may be so limited as not to warrant hardcopy at all—except on demand. Future contracts might read this way: "The publisher warrants, through the announcement and promotion of Professor X's work, that it is bona fide and measures up to the highest standards in the field of Y. This warranty is based on expert referee reports, expert in house editing and expert capacity to reach those significant others who might be enticed or induced into purchasing a manuscript."

Such a situation is not imminent. But we must face the implications of the division of labor not of publishers' choosing—that has signified the entrance of new players into our world. These players will at times be highly abrasive and competitive. They will not go easily into the night. Some of them may be absorbed by large publishing houses, but some of these large publishing houses may themselves be absorbed by the new technology.

These battles will be fought out at the top of the market, not among niche publishers. Publishers, like universities, will become increasingly differentiated as to their offerings and expertise. A house such as Transaction, with a network of people built over more than thirty years in a variety of fields, might well fare better than omnibus firms attempting to publish everything in every field and also better than university presses that feel the need to issue everything from texts in microbiology to pamphlets on modern poetry.

The new technology is more likely to respond to niche publishing as a source than as an objective for takeover. In any case, niche publishers have little to fear from electronic publishers, who do not have the capacity to do what traditional publishers do well: sort out quality from quantity. True publishers generate quality and evaluation of manuscripts. Houses that emphasize reproduction of images at low returns are badly behind the times.

The contracts that small and midsized publishers should develop with document delivery services are analogous to those they form with overseas distributors who are better placed to deliver to foreign marketplaces than the principals they represent. The division of labor is part and parcel of the evolution of the marketplace. This movement toward hard specialization is something to be monitored, not feared; understood, not competed with.

In their haste to comply, publishers must be careful not to be suckered into easy responses to questionnaires which somehow presume that, if we cannot place a checkmark in the appropriate box, we have somehow failed in our capacity to keep up with the McGraws. Our essential, ongoing need is to keep up with scholarship, to maintain a sense of the cutting edge of research, to appreciate the diverse needs of teachers and students on a particular subject. If we do this, scholarly publishing will do just fine, whatever happens in the volatile world of product innovation, computer technology and telecommunication systems.

Bringing scholarly products to the market has now become a matter of months, even weeks, instead of years. Publishers will have to meet the challenge of a new sense of time no less than the reduction of space in the world of information. At the same time, the electronic document delivery people will have to appreciate better that their bread and butter depends upon demand, not for raw product, but for good articles and good books.

Universities will still have to produce good ideas, and the professional publishers will have to bring such ideas to the market. In the drawing together of the marketplace of ideas and the marketplace of commerce, we will enter a new millennium more in hope than in trembling. The division of labor will be our savior. The attempt to do everything would be our doom.

7

Social Science and Scholarly Communication

Thinking of the past, my thoughts turn to all that I have learned and gained from having known individuals like Jeremiah Kaplan, founder of The Free Press, Irving Kristol and Martin Kessler, major figures in the start up of Basic Books, the too often neglected Frederick Praeger who pioneered two start-ups (one that still goes by his name, the other being Westview Publishers), and my good friends from my Buenos Aires days, Enrique Butelman, who founded Editorial Paidos and Jorge Grisetti who co-founded Siglo Veinte Uno. In this astonishing select group, Hans Zetterberg also figures prominently; for in the decade of the 1960s, through the firm of Bedminster Press, he issued a series of classic and contemporary studies that remain part of the social science bedrock. Hans understood early on that social science is a business, and business is not a bad word subject only to mindless ridicule by the ill informed.

So here I stand on the shoulders of these giants, and untold others, in demonstrating the possibilities of, and the need for, a social scientific mission in the world of professional and academic publishing. It is no simple task to have one foot in the world of university press-type books and journals, and the other in the still riskier world of commercial publishing. Add to that mix a continuing effort to contribute my own scholarship independent of publishing, and the mix can easily become quite a smorgasbord. In two years marks the fiftieth anniversary of Transaction, an organization that I have been with since its start, so my impulse to reflection and rumination is, I hope, understandable. It has been my good fortune that the world has conspired to bring scholarly communication and social science closer to each other than ever before.

This is the case not simply because niche publishing in the area of social research has been virtually abandoned by the giant commercial houses. Nor is it because a new cluster of specialist publishers has come into existence to address such concerns; it is rather that policy issues have arisen that now require urgent attention. Just think of the plethora of issues we are now faced with that were virtually unknown even a half-century ago, when at least some of us embarked on careers in social science. There is the relationship between physical property and intellectual property; private ownership of information and the public right to know; national boundaries and international space; the emergence of English as the *lingua franca* in global communications amidst demands for preserving national and regional cultures; and huge shifts in how information is transmitted to the public, from hard copy, to broadcasting, to the Internet. Thus it is that those of us involved in publishing, far from being odd ducks or adventurers, are in fact at the cutting edge of social research questions. What started as the accidental choice of a few has now been transformed into the necessary agenda of many.

Both social science and scholarly publishing must break the barriers of parochial boundaries. From the unified nature of international development to the integrative capacities of the Internet, internationalism is the order of the day. This does not signify a loss of sovereignty, of family, or even of individuality. It most assuredly is not a call for imposition of ludicrous notions of world government and world order, or other such dangerous forms of control. It is recognition that knowledge and communication as such break down every barrier. Social science in this admittedly highly confined realm is part of the democratizing process. In focusing on stratification and inequalities, on different paths to pursue universal aims of health, education, and security, the social sciences have been transformed. It is not so much placing limits on the range of theorizing as seeking avenues of performance that actually assist the common good. For example, the International

Social Security Association located in Switzerland is the center of worldwide efforts to reconfigure social security in every nation in the democratic West. The European experiences become the source of intelligence in establishing new directions in the United States.

We are all the beneficiaries of global experience on concerns that are paramount for each of us as individuals. We live in a world where at the click of a mouse we can access the news and ideas of literally thousands of significant publications worldwide. In such a communication environment, of what use or meaning are publications such as the *American Sociologist,* the *Australian Journal of Anthropology,* or the *Indian Economic Review*? If the social sciences are to make credible claims of universality, they cannot simply be reduced to locality. Of course there is sociology of America, and an economy of India, and anthropology in Australia. But these are aspects of the universal claims for the scientific parameters of sociology, economy and anthropology not the special properties of individual nations. The attachment of the names of specific nations to intellectual pursuits has become gratuitous at the least and insulting at worst. Parochial labels create artificial boundaries on what can be studied, and isolationism in science, as in politics, is counterproductive.

Open communication channels and well-developed social science paradigms are part of the world of knowledge. They are not intended to be means to political ends and whims. One might make a case that communication and social science extend the boundaries of democracy, but they do so precisely to the extent that they are not subject to political and bureaucratic pressures. A great failing in much of current social science and communication thinking is the idea of putting such knowledge at the service of elites, of political or economic agencies who have little interest in the general will or the general good. That is why social science knowledge is an end in itself, not a means to securing the interests of others. The perfume of power emits a strong but transitory odor. The tasks of the present are to make our respective professional associations responsive to the citizens we serve and the membership we aim to enhance. When such organizations become footnotes to political special interests, the very claims to science become suspect and corrupted. There are indeed risks in converting our areas of expertise into a "new class." And the goals of serving a constituency beyond us are not easy to establish. But such risks are at least manageable and controllable by us. To tailor the findings and theories that we have in the service of unknown ends is, in my view, the more risky alternative.

In discussions of this sort within the social scientific communities, the emphasis is too often on the social and not the scientific. In point of fact, it is precisely at the level of the sciences that scholarly publishing and academic social research link up. For the very act of reading, like that of research, involves forms of rationality and reflection that helped maintain, if not bring about, the very existence of civilization as such. In the present epoch, the nature of this struggle has been clear by the coming to power of regimes that first burn and then ban books. It is also a struggle to ensure a multiplicity and not a monopoly of publishing. Without competition in the marketplace of ideas the need to burn books becomes superfluous. The fascist approach is more obviously brutal. The communist approach is both more sophisticated and more effective. The rise to academic respectability of doctrines that emphasize irrationality, subjectivity, and ideology as substitutes for research and experimentation must also be understood as a frontal assault against the social sciences no less than the cultural heritage. At the start of the twentieth century, Charles Sanders Peirce in his essay on "The Fixation of Belief" somewhat whimsically declared that 99 percent of the educated public had their views "fixed" by tradition and authority, and about 1 percent by the methods of science. Whatever those numbers are in actuality, it is our shared and often lonely task to stand with science as a method and exact information as a universal goal. This struggle is in effect the core of our politics. We must make a constant decision to oppose those who seek the triumph of the Will against the pursuit of learning. Doing so is also a tacit decision on behalf of living in a world in which truth is steadily pursued, and shortcomings necessarily tolerated, not gleefully punished.

Scholarly publishing and social science research fuse in another important respect. We are constantly involved in refining and redefining the tradition, the canon if you will, of intellectual life. The publishing community is compelled to do so by keeping a close ear to the ground listening to the demand curve of the marketplace, to learn what people want to read, and what they will buy for themselves or their libraries. What is transient and what is permanent is less a series of guesses than a statistical series of tables of buying habits. The social science community for its part is largely responsible for the supply curve—what it is that scholars are working on, what is exciting in the combining and recombining of the universe. And here scholarship is at the cutting edge, and good publishing follows closely behind. In the past, publishing carved up the world among commercial, text, and scholarly modalities. But that approach is probably as archaic and outmoded as those who still think of social research as being encapsulated by the fields of anthropology, political science, psychology, sociology, and economics or those fields, which saw themselves, organized in departmental terms between 1896-1912. The real world ultimately gets what it wants and discards the rest. This is a harsh reality, but one that is disregarded at the peril of science and society alike. A wide range of fields such as criminology, demography, urbanology, communication studies, cognitive studies, social statistics policy research, and decision theory sprang up and contributed to the size and potency of social science in the twentieth century. These are also the fields that have prevented hardening of the arteries in scholarly communication.

These brief remarks hardly exhaust an examination of Social Science and Scholarly Communication in the New Millennium. Each has power requirements that have precious little to do with each other. The marketplace of supply and demand and the marketplace of ideas are hardly isomorphic or symmetrical. Losses of profitability at one end, and failure to reach out for innovative manuscripts at the other are a consequence of risks. That said I believe it is of utmost importance to see the forces at work that have brought us to this point, a point of synthesis unforeseen in past ages. Researchers and publishers are united by the act of creation. Gangsters and terrorists are united by the act of destruction. If this strikes a somewhat melodramatic, indeed avuncular, note on which to end, I have done so with a purpose—to emphasize as strongly as possible that ours is a world of hard struggles. These are our struggles. They take place in the very bowels of our lives as social scientists and academic publishers. It would be a forlorn task to go hunting for the perfect political party to represent our interests. We need to define and defend our interests in terms of our quotidian lives, and in so doing make certain that the political processes respond to our needs or risk our wrath. To resort to an old and tired cliché, we live in times of enormous upheaval and uncertainty—but so did all those who came before us that engaged in the struggle for a decent world filled with decent people.

Two caveats of deep consequence need to be registered. First, my position is not aimed at recreating an adversarial model of earlier times in social science history. Whoever comes before our bar with clean hands and a simple need to know should be treated fairly and with respect. Whether the purchaser of our wares are economic lords, political pashas, or labor barons, or for that matter, representatives of those seeking a greater share of equity in the larger society, such as welfare recipients, black male prison inmates, or groups with physical or psychological handicaps—they should be treated with respect. They should also be told in no uncertain terms: *caveat emptor.* The "buyer" of social science goods or services may or may not get what he or she wants, or expects.

In addition, as a second item of importance: the fact that the social science communities seek to develop a solid phalanx to represent its interests does not mean the end of debate and discussion among social scientists as to the rights and wrongs of a particular position within theory, methods, policies, or outlooks. The existence of powerful interest professionals in medicine, law, or for that matter, biology or physics does not demarcate the end of debate. Sharing a common guild, union, or profession is a recognition of great maturity, that is to say, that a group of diverse individuals with similar training and background can get together to assert common and shared goals about livelihood and ultimate purpose. Far from spelling the end of debate or discourse, it could well stimulate such debates—since the appeal will be to represent specific views to the larger societies and communities. I take this to be the full and rich meaning of Hans Zetterberg's assertions in *Social Science and Social Policy* that having a research proposal for sale is not much different than having a product for sale. The buyers of our properties; material or intellectual, rich or poor, mainstream or marginal groups, uplifted citizens or downtrodden souls, will readily come to understand the difference between snake oil and social science finding.

This brings me to a final, consequential point: interests are collective and organizational, but impulses are individual and private. Ultimately, it is as individuals that we need to develop the resolute skills to defend an honest social science within the context of a free communications system. In *The Plague,* Albert Camus put these sentiments best for me: "There always comes a time in history when the person who dares to say that two plus two equals four is punished with death. And the issue is not what reward or what punishment will be the outcome of that reasoning. The issue is whether or not two plus two equals four."

8

Open Access and Closed Minds

There is an old Leninist adage, invented by Lenin himself of course, that if the revolution supplies enough rope to the bourgeois class, it will manage to hang itself. This cynical aphorism, born of a belief that class animosity has a silver lining known as utopia, is a solid tactical formula for success. Whatever its failures or successes in other fields, publishing has become a fertile ground for such practices—and theorizing.

The latest craze for this approach stimulated by the noble purpose of spreading the world of knowledge to all at the lowest or no cost goes under the banner of "open access." This signifies quite simply treating intellectual property as little more than a drink of water—a right to the recipient and an obligation by the creator. It is intriguing that a number of large publishing houses have latched on to this pseudo-democratic formula, more in fear of offending library purchasers than any ideological commitment to self destruction. It seems to enhance the democratic process—one that promises availability of information and knowledge at no cost.

The disastrous history of newspaper and magazine publishing in the first decade of the twenty-first century seemingly made no impression on the barons of the printed word. The conversion of most major cities into single newspaper towns, and sometimes not even that, has not served to alert this sector of the publishing world, only to have developed devious methods to be required for supporting "open access" by advocating such mechanisms as pre-payments either by authors or by institutions of higher learning to ante up heavy funds for articles and monographs that were indeed often readily available for minimum funds. Hence, instead of purchasing a journal with the desired article for a relative pittance, the author or the agency must now ante up thousands of dollars as a precondition to permit "open access."

This bizarre formula allows the not inconsequential social welfare sector of the library community to continue to wave the banner of open access for free minds, while enlisting the support of major firms who see this as a mechanism for reducing costs of publication and shifting the burdens from the marketing place, as such, to the authors sometimes desperate for a public hearing, or to their universities, largely oblivious to the back door payout required to gain access to anything but information open to all.

The World Library and Information Congress, scheduled for August 2010, and to be held in Gothenburg, Sweden, may fall under the conventional radar screen of other members in the communications area, so it merits a close examination. It is vigorously dedicated to "open access to knowledge" and "promoting sustainable progress." There is no question in the ideology of the sponsors that this is a "motto" not a query. We are further informed that this is a libraries initiative. We are told that "free access to knowledge is as important as freedom of speech"(although many sectors in this library world have remained silence in the face of challenges to free speech in places ranging from Cuba to Iran to China). We are reminded that open access means "accessible for all, even for the visually impaired or others with reading difficulties."

Lest the concerns of this sponsoring agency still seem unclear, the program goes on to state plainly that open access is inclusive, open to all "no matter whom you are or where you come from." Open access further claims to be essential to "public domain, a place on the net or in a physical space." It is also a place where "people contribute to and socially share content produced and owned as a public service." Indeed it is a place "where the user might as well be the producer." The world of imagination is one which there is a "better balance between copyright laws and freedom of information." We are assured by the sponsors that "progress" for one individual is progress for the society. Needless to add, the content of such open access publications are not subject to prior examination. To do so would only serve to impinge on the national sovereignty of others.

The socialist character of this manifesto is hardly restrained and to the credit of its library sponsors, not disguised. There is little patience with the idea of proprietary considerations, differences of opinion as expressed through the marketplace of ideas, and none whatsoever to how such a manifesto is to be brought into play without cost, pain or preference. The list of sponsors prompts one to be assured that with publishing houses and information services as De Gruyter, Gale, Elsevier, Oxford University Press, Sage, Springer, Thompson Reuters, among others, the line between those hung and those doing the hanging is transparent.

The plain fact is that the Open Access "movement" is the long awaited sequel to the Fair Use "doctrine" that became the gaping hole in the revised copyright legislation of 1975-76. How an exemption became a battering ram for end running the major provisions protecting intellectual property is a story for another time and place. What is a story for this time and place are the tragic consequences of a rupture between two essential communities serving the publishing world: librarians and the publishers themselves.

The struggle erupted as a specific condition of the larger battle between socialist and free market concepts of the world of information. For their part, the librarians—by and large—came to perceive of communication as an inherent free commodity, not only in its circulation, but in its very production. For the publishers—again by and large—information like entertainment is defined as useful by what a larger public buys and pays for. In this contest shadowboxing with the issues became the norm.

This bitter rift, conducted under conditions of outward organizational calm and professional civility, could have readily been avoided were the publishing community at the very onset of new rules and regulations governing intellectual property saw fit to include—rather than exclude—libraries from a share in the profits of a new legal construct that protects personal creation as proprietary in its nature, just as patent rights and technical inventions are.

The venality of monopoly publishers thus entered the fray with the ideology of librarians—to the benefit of neither and detriment of all. The inability of either side to see the consequences of this rift prevented the formation of a decent public policy. The courts of law had a field day deciphering and settling differences to the benefit of large legal firms, and the detriment of the information network as such. It might well be argued that the stakes were too high and the rifts too wide to have permitted compromise and consensus. But that is a conjecture that can hardly be proven, much less speculated upon.

What is beyond conjecture is that neither librarians nor publishers have surrendered their distinctive claims. The former holds global gatherings organized around ideological premises of open access, while the monopoly publishers join powerful librarian agencies at such gatherings, for the scarcely designed purpose of subverting the very reforms that librarians clamor for. Fair use became quantified in terms of copies that can be produced in conditions that are strictly limited. As a result, instead of an exemption, it became an impediment. Needless to say, the notion of fair use became a legal battlefield—one fought out by surrogates: major corporations protecting their investments versus public interest groups in which the public vanished as a concern for *vox populi* as such.

Monopoly publishing is linked to corporate activities in the media industry as a whole. Arguably, the book and journal industry has quietly learned from the television field that information is not free, but quite costly. To start with, every intermediary delivery system from Google to the giant national networks such as CBS, NBC, ABC, and FOX derived huge revenues from advertisements—both direct and indirect. In addition, the television viewers must pay blanket "basic fees" for programs rarely if ever watched. Finally, the system is so structured that the very appearance of images depends not on viewer discretion but upon broadcast revenues. In light of such a situation, it might well be that viewers would greatly prefer paying for direct feeds rather than for the mythology that viewing is without a cost to consumers. The illusion of freedom of information is fostered by the absence of direct cash outlays as in the world of hardcopy books and journals. It is a dangerous pattern to adopt, and one that profoundly shifts the burdens from consumer preferences to corporate profits. The actual realities knock out the props from under the "free access" model more profoundly than a barrel of slogans.

If ever there is an area in which public policy would have been of great benefit, this would have been the case with fair use. That area more or less circumscribed and repeatedly violated is no longer on the agenda. In its place "open access" has become the rallying cry. And once again, the publishers have interpreted this demand for recognition of a public right to know as taking place and being strictly circumscribed within the private right to profits. The same monopoly publishers who can be found parading about at such gatherings of librarians showing off their new gadgets and systems, do so not to celebrate open access, but to remind the librarians that open access does not mean free access.

The major players in journal and book manufacturing and distribution now argue, belatedly and in draconian fashion that they too support open access. They simply want the authors, or their surrogates, universities and research facilities to underwrite the cost of such openness. For a set price—usually in the thousands—a scholarly article can be open to all upon payment of a considerable sum of money. Needless to say such major firms do not much care where the funds come from, as long as the cash flow is unabated.

For their part, the librarians have developed their own form of myopia, acting as if demands for open access are a prelude to free access. Socialist ideology dictates that if water is free, so too should this be the case for information. Ignored in all this fantasizing is that water may be free in the ocean, if one can overcome the waves and defy the sharks, but that in its distillation and distribution to human communities, waterworks are a costly business—usually paid for by taxpayer funds. The more learned members of this community recognize this fact, and argue ably that information should be paid for only once, not twice by public funds, or at least funds determined and determined by local, state and federal governments, or by large universities in which research is reputedly being conducted.

The problem with such equations of research monographs with books as they finally appear is a total discounting of the division of labor that exists in this complex mosaic of writers, researchers, funding agencies and other relevant sponsors. To start with, publishers, by providing direct support as advances and indirect support though royalties are a vital link in the well being of the property creators. The publisher also provides the bulk of their support to a book *after* publication—through marketing, advertising, promotion, warehousing, designing, artwork, and global networking in support of distribution, reviewing and translating. The alternative is not socialism, but marketplace anarchism.

The issue dividing publishers and librarians is not a passion for an open and free exchange of ideas; but the disintegration of standards—already under siege by extremists from all quarters. So many online only journal activities have essentially by-passed the referee-review process. And that means every reader takes pot luck and every scribbler feels that any statement (about people, places or events) he or she makes is equal or greater in weight than those made by others. As a consequence the carefully and imperfectly stitched networking of ideas built up over the centuries in universities and institutes of learning become victims in this process of the pure relativizing of ideas and disparagement of even the idea of standards.

This battleground of ideas with many potential victims and few victors offers succor only to the legal profession, who in the absence of moral standards or guideposts, are left to sort out who gets what and how, when, and why. Earlier this year in March, the initially publisher sponsored agency the Copyright Clearance Center held gatherings in which lawyers proclaimed the new rules of the publishing game, while nary a publisher was found on panels having to do with their own rights, or for that matter, library responsibilities. This economically unwieldy and ethically unhealthy condition will continue for as long as the communities directly charged with responsibilities for the steady and free flow of information continue to shadowbox and disguise the ideological sources of such madness in the "drink of water" theory of information, data and ideas.

Information and ideas of larger consequence will continue to flow—of that I am certain—but the critical issue is who exactly is to pay for such a flow? There are risks in monopolization of publishing, in which a dozen houses now account for 65 percent of the revenues according to the most recent Book Publishing Industry Report. The top six (Pearson, Houghton-Mifflin, McGraw-Hill, Cengage Learning, Scholastic, and Bertelsmann) account for 52 percent, while the second tier six (News Corp, CBS, Hachette, Readers Digest, John Wiley, and Reed Elsevier) account for a little more than 13 percent. The other thousands of publishers who bulk up the Frankfurt Annual Catalogues divide up the remaining one third in fractional terms.

But one dozen goliaths and several thousand Davids look mighty tempting in a world being proposed through "open access" formulas, in which the government alone will determine what is open and what is closed. The recent actions by the governments of Iran, Cuba, and China not only to old line publishers but new line electronic servers such as Google, should provide a sobering warning to those devoted to the open access model being served up as a democratic option. The secret in the closet is "institutional support"; or more directly, shifting the burdens of payment for services from the private sector to the academic sector. How long will it be before demands for refereeing services become a cascade of noises from those who now advocate "open access." It will prove chimerical in the extreme: neither open in information terms, nor accessible to the larger public that continually is in need of fresh ideas for a free society.

Part 2

The Political Economy of Publishing

9

Professional Ambitions and Public Interests

There were few more penetrating commentators on professional life in America than the late Everett Cherrington Hughes. In a series of articles, capped by one in the *American Psychologist* nearly sixty years ago, in 1952, he discussed the relationship between science and profession and their impact on psychology. His observations remain central to the current agenda of social research, and they merit direct quotation: "The question of competence is discussed in complete separation from the outcome for the client. In protecting the reputation of the profession and the professional from just criticism, and in protecting the client from incompetent members of the profession, secrecy can scarcely be avoided. Secrecy and institutional sanctions thus arise in the profession as they do not in the pure science."

Hughes was not launching a broadside against professional life so much as noting that its imperatives and directives depart substantially from the norms of scientific method. In his time, he fought mightily to expand the responsibility of professions to public opinion. He was in favor of increasing the scope and size of occupations, thereby broadening the opportunity for participation. The problem became that these occupations gave rise to new professions. And while this plurality, this multiplicity, accorded with Hughes' sense of the free and open marketplace, he could hardly anticipate the next step in the evolution of professionalism in America: the developing effort of the professions to serve as imperial vanguards for the "science" they were supposedly serving.

This is the current focus of struggle between scientific research and professional organization—whether to service the professional or to serve the public. Changes in the life of the academy will occur rapidly, or not at all, depending on what takes place in the field of scientific struggle.

The lacuna in Hughes' analysis was his failure to see that the struggle over the nature of science in society takes place largely within professional life. It does not emerge simply as some antinomial relationship with occupational roles. Each profession reflects a discipline with a unique inner history. In economics the struggle is between monetarist and institutionalist claims about reaching the nirvana of equilibrium. In sociology the struggle is between functional-positivist and cultural-subjectivist modalities. In anthropology it is a struggle between the traditional notion of empirical work in the field and those for whom the field is no longer overseas, but occurs rather in the forces of oppression at home.

In political science the gulf between normative theorizing and empirical description could hardly be wider. In psychology the key rift, indeed rupture, is between clinical uncovery and experimental discovery as methods for determining evidence. In short, professional life is closely allied to pivotal conflicts and movements within each of the social science professions. Hughes saw largely external forms of combat, those in which the administrative apparatus of professions act preemptively, not infrequently behind the backs of its own membership. But the play of forces on a professional level involves interior dynamics that must be identified by those who want to understand the superstructure of social science.

Hughes certainly appreciated that for better or worse we share an open society premised on free enterprise. Some professional organizations make no bones as to their own preference for planned and regimented social systems. Especially now, as the communication of ideas, facts, and information becomes not only a competition for attention but also a distribution of influence, the defensiveness of leadership in professional organizations is all the more appalling. In such an ideological climate, it is inevitable that these organizations seek to dominate not only membership scrolls, but also struggle to control communication outlets of the sciences. And that is where the issue is joined on the ground—between the pursuit of science and the self-interest of professions.

The character of social science changed dramatically in the past century. From the traditional "big five" emerged a much larger group of disciplinary activities: communication studies, demography, urban affairs, policy and evaluation research, criminology, and many others. Social theory is common to all social science, new and old. Methodology is equally common to all of the social and behavioral sciences. The concept of a "general theory," or an imperial vision of any single social science over all others, is a useful fiction whose end has clearly arrived. Even a cursory review of such attempts also suggests a futility about what a post-imperial vision would look like. But the imperial vision was never simply a catching up to modernity; it was born of a deeply conservative impulse to discover what makes life meaningful and worth living. In its transformation into a variety of utopian meandering, the theories of totality have finally given way to more modest endeavors that seek to give an account of and to predict social change. This is why it is all the more disconcerting that professional organizations have failed to learn the lessons of failed theorizing on the grand scale. Such organizations have become empires themselves.

The very effort to reach out to a larger public compels breaking the boundaries of disciplines set in abstract categories. Theories do not confirm each other; they are embedded in the empirical realities of an age. This is a technological era in which the social sciences are driven by issues and events, not ideologies—by concrete human and global needs rather than inherited systems. This is not to say that the social sciences are now Balkanized, positivistic, or immune to general theory. It is to say that the basic mission of the social sciences is now more clearly demarcated and delineated than it was in the past. The character of scholarly publishing in this era demonstrates a marked decrease in abstract theory and a corresponding expansion of studies in applied disciplines – from crime control to health expansion.

It is dangerous for old doctrines to claim superior wisdom over and against new formulations of a more modest sort. We find such self-serving claims gratuitous and self-defeating even if in the short term they appear to be self-sustaining. We can recognize the achievements of the past without developing inhibiting icons that presume decay in the professional culture. It is true enough that the parameters of social science have changed significantly in the past fifty years—and they doubtless will in the next fifty as well. It is always better to avoid ethical labels and focus instead on identifying and developing the best of the present. Publishing must face the fact that it is enveloped more by specific public needs than by general insights by presumably great people.

It is also essential to ensure that publications help scholars, younger ones in particular, think outside the box in which they were presumably trained. Associations are jealous, and often un-forgiving, animals. They want to serve as the reservoir of all learning, and to convince their members that all truth resides within a single mansion. The alarming tendency of professional societies to place their own stamp on society's publications is at the same time an effort to delegitimate publications beyond their direct control. This has become a critical struggle of the moment, for professional associations today are far more insidious and powerful than they were forty years ago. They give rewards and convey status, and they do not wish others to do so. Although there are a plethora of journals in each of the social sciences, those publications that carry the stamp of officialdom are generally favored with respect to tenure and promotion.

That there is a broad recognition of the risks and dangers of professional imperialism is made evident by the rise of new schol-arly groups in psychology, history, economics, and culture, all of which have emerged, from older disciplines. Beyond that, we need to factor in the rise of new disciplines whose fund of knowledge and inspiration comes mainly from the social sciences. The path of democratic struggle is not a perverse insistence on a particular interpretation of events, as much as openness to new policies and innovations. These new disciplines, often of an applied nature, serves as an exemplar to general forces of academic and policy life that advance the open society.

There are essentially three ways in which a better balance between professional desires and public needs can be brought into some sort of equilibrium. In each instance, the courage and fortitude that individuals have had in the past, or will have in the future, will make this worthy outcome take place.

First, new fields of research and theory, such as those already mentioned, continue to evolve. Such new combinations and permutations will prevent old-line societies from hardening of the arteries. Areas such as criminology at one end and communication at another, boast multiplied organizational formations. Even these new fields are "twigging" in order to satisfy specific needs and demands of a social nature. Forty years ago this field was nascent, and part of sociology or psychology departments. The explosion of social science knowledge has brought about associations that dwarf in size, and even in importance, older academic formations. They do so to such an extent, that the very paradigms of social research bequeathed to the twenty-first century are properly subject to questioning.

Second, a consequence of this knowledge explosion has been a parallel emergence of new journals. These enlist a whole new cluster of editors and contributors, taking place throughout policy organizations and university departments in Europe and Asia no less than North America. Such journals often acquire a status and importance to specialist audiences that dwarfs the stifling conformity of old-line journals. Indeed, in international relations, journals such as *Foreign Affairs, Foreign Policy, Orbis,* and the *National Interest* easily outshine in quality and outdistance in influence and in terms of professional significance, publications of academic organizations grounded in international and regional associations.

Third, even within old-line organizations a "twigging effect" is taking place. In areas as distinct as psychology, history, and cultural studies, new societies have emerged to challenge—sometimes directly, other times elliptically established organizations. Indeed, the five major organizations that represent anthropology, political science, sociology, economics, and psychology were all formed roughly one hundred years ago, at least in the United States. The wonder of it all is how these big five have managed to hold sway for as long as they have. But again, in basic social science fields, reality has overtaken sloth to the point that professional life can ill afford to ignore, much less disparage larger public concerns. Such fields as evaluation and policy, criminology and penology, communication and information, are not simply new national structures but international associations. These also challenge the parochial and insular quality of established traditional societies.

To speak of the parochial and insular character of older professional societies takes us into the realm of ideology and utopia. In violation of the Weberian distinction between the calling of science and the conduct of politics, too many of these old-line agencies act as if proclamations about the desirable are the same as statements about the actual. So we are offered hortatory proclamations on race, nationality, and sexuality ranging from demands for statehood of the District of Columbia, opposition to military intervention in Iraq, Afghanistan, and elsewhere, to assurances of new standards as to what constitutes the nuclear family. Some of these proposals may prove to be feasible political goals. But instead of opening up discourse concerning vital issues, they have the effect of closing down discussions. Instead of the scientific examination of abrasive social problems, we are offered smooth-talking palliatives that can barely be taken seriously as analytic claims, and even less seriously as professional platforms. Such tendentious postures serve only the cause of alienation from professional life and cynicism about the scientific calling.

This is not a call to arms, nor an effort to undo all the good that professional life has accomplished over the years in creating a scientific base and public trust in the work of social researchers. It is a call to attention that knowledge is organizationally defined and promoted. New research organizations have emerged from one end of the land to the other, and new centers of authority in universities formerly not considered part of the avant-garde have emerged. Not only does private initiative or private imagination assures continuity, but larger public developments are an integral part to our human efforts.

If this cursory review of a singular history of a single publisher, Transaction, lacks the sort of elegiac note that might have been expected on so special an occasion, it is simply that self-satisfaction is a risky bit of emotional baggage to parade forth when so much is at stake and so much more still needs to be done. General theory may be in disrepute, but subjectivist anti-theory is a medication far worse than the disease. The vineyards are many; the field workers remain few. The triumph of social science and democracy over the forces of darkness is by no means assured in the book of life. We still require human intelligence and imagination to make good on such a bold vision.

I opened with an observation by Everett Hughes, so closing with a statement from Robert Lynd's *Knowledge for What?* seem appropriate: "Social science is confined neither to practical politics nor to things whose practicality is demonstrable this afternoon or tomorrow morning. Nor is its role merely to stand by, describe, and generalize, like a seismologist watching a volcano. There is no other agency in our culture whose role is to ask long range and if need be, abruptly irreverent questions of our democratic institutions, and to follow these questions with research and the systematic charting of the way ahead. The responsibility is to keep everlastingly challenging the present with the question: What is it that we human beings want and what things would have to be done, in what ways and in what sequence, in order to change the present so as to achieve it?" This is not a rejection of large scale theorizing, it is a rejection of the social sciences ignoring the realities and limitations of history and culture.

10

Formatting Ideology through Tabloid Politics

I

One of the residual consequences of the "decade of the 1960s" for the new century is the demand of academics in the liberal arts for the political relevance of what they do. The oft-cited biases in the media have their roots in that decade, and have consequences in the present as well. The tactics of the past have been incorporated in the calling of the present in a variety of ways: mass campus demonstrations, civil disobedience and even disruption, and the confessed use of the classroom as a forum for class struggle. The issues have changed, environmentalism, health care, and sexual liberation, among other agenda items. But the most lasting impact of the past has been the idea that to be relevant is to be absolutely dedicated to blind partisanship. It is clear that such demands invite cynicism from the larger public less concerned with academic labels than public trusts.

This shift from reporting to advocating amounts to an admission that bias rules the mind, hence the issue is bias in the service of specific interests, causes, or movements. But how are such ambitious goals translated into academic performance? An initial answer was coincident with the establishment of the *New York Review of Books* in 1963. Its use of a tabloid format excited the imaginations and of top figures in both the liberal arts and liberal politics, and helped the editors enlist their support. What began opportunistically as a publication to fill empty kiosks during the 1963 strike of the *New York Times* has evolved over the years into a publication with a full-blown style of its own. This highbrow tabloid, filled with reviews, has become a bridge into mass culture, or at least the feeling of mass culture, for academics in liberal arts departments in institutions of higher education and those who look to them for intellectual direction. The *New York Review of Books* success may also help explain the deadening uniformity of opinion evident among the chattering classes and their organs of opinion.

What is different at present is the flowering of the conservative response to the decade of the sixties. Indeed, the conservatives have developed magazines, journals, and book publishing outlets that have created a parallel set of forces. In this, Fox News Network in all likelihood is the quality of, if not greater, than the pulling power of the liberal "elite" publications. Conservatism has established an elite of its own. What one sees as American enters the second decade of the century is a counter-trend that relies on a mass base that is more located in geography than in ideology.

The tabloid format has enormous advantages over other mechanisms of political protest available to those in academic life. It conveys a sense of journalistic urgency, and potentially has outreach and impact far beyond publications presented in conventional journal formats, especially those that are viewed as elitist publications, such as the *Virginia Quarterly Review,* the *Massachusetts Quarterly,* the *Michigan Quarterly,* the *Sewanee Review* and other publications rooted in English departments at major state universities throughout the United States. The tabloid format adopted by the *NYRB* also broke the dominance of well-established political types of liberal arts publications such as *Partisan Review, Hudson Review,* and *Kenyon Review.* Indeed, the *NYRB* joyfully raided these earlier efforts at relevance for prospective authors ready to participate in a new revolution of sensibility institutionalized in the 1960s. The weakening and in instances the closure of some of the liberal anti-communist publications of the 1930s may have been the most obvious victim, of the liberal pro-communist (or at least social welfare) publications of the 1960s. It is a strange rhetoric to link publications with time frames and cultural generalities, but it is the cultural superstructure with which publications must deal.

A model for development of such a cultural periodical already existed: the *Times Literary Supplement* (*TLS*), which had already gained a foothold in Anglo-American culture as the leading public arbiter of the culture. Books reviewed favorably in that amazing publication, having already past the celebration of its one hundred years of well-deserved preeminence, meant a great deal to the author, his or her publisher, and the cultivated classes. But for the new partisans of ideological fervor, the *TLS*'s aloof, almost disdainful, disregard for predictable and predetermined outcomes to its review process (and the editorial process in general) was not a virtue, but a vice. One might argue that the *TLS* thrived precisely because of its indifference to fashion and the lack of predictability in its choice of books reviewed and reviewers chosen (despite favoring books in political history and personal biography). But for those with a passion for directed change, this core aspect of the *TLS* tradition represented a window of opportunity for new activists with new publications in the United Kingdom, including the *London Review of Books*, and the *New Left Review*.

Into the breach stepped liberal arts academics with a mission and a tool: the intellectual tabloid. It was an answer to a prayer: a medium based on broad popular appeal (it was derived from the penny newspaper), liberated from the taint of "elitism," and printed on cheap newsprint. As it evolved, the only fly in the ointment became the word "books" in the publication's title, pioneered some forty years ago. No matter how small the type, the words "of books" remain there for all to see. Over time, "books" came to denote something less compelling than the mass media that have captured the minds and hearts of the people the publication was intended to reach.

Within these limitations, the *NYRB* has retained its mission of partisanship over the course of the years. Initially, books were reviewed as part of exceptionally long commentaries rather than traditional reviews. Next, the reviews became review essays, an artful assemblage of titles that would permit the famous and the quasi-famous to make their statements. Finally, the wraps came off, and increasingly, the articles lost any pretense of being reviews, and were simply articles and extended op-ed pieces on topics ranging from political leaders to military affairs. The Vietnam War provided the trigger for this new approach.

But it was more than foreign policy as such that concerned the editors of *NYRB*. It was the struggle for truth, as represented by the Democratic Party, against error, as represented by the Republican Party. Through Richard Nixon, Ronald Reagan, and the two George Bush presidencies, the publication increasingly defined its mission as the salvation of the Republic by novelists, literary critics, editorial page commentators, historians, for whom it would provide a forum to express their outrage and indignation against the course of Empire. And if one is to judge its success in achieving this mission by examining the political leanings of liberal arts departments at colleges and universities—the place at which the publication reaches its core audience—the editors have succeeded. Political debate in these core audiences centers less on the full range of political options, and more on strategies and tactics of Democratic Party responses to the Republican threat. That *NYRB* is in the vortex of this undeviating consensus is beyond question.

II

The *New York Review of Books* has harnessed the normal anti-political bias of the liberal classes into a frontal assault on legitimacy and leadership as such. While the biases are clearly heavily weighted against the Republican Party, over the years, the *NYRB* has taken its swipes against the Democrats as well—especially Lyndon Johnson for his pursuit of the Vietnam War. While milder in tone and more in the nature of rebuke than frothing assault, the driving force seems to be revulsion with the system, the nation, and pardon the expression, the regime, that dominates certain sectors of American life. In this sense, the *NYRB* has become something of a European cultural outpost rather than a nativist rebuke to the dominant culture. As such, its seeming linkage to Democratic Party concerns is more apparent than real, reduced to the academic rump of liberal arts departments rather than mainstream political party life as such.

The *NYRB* approach to the Iraq War well reflected this. Its assault on Imperial America, so often heard before and during the war against Saddam in Germany, France, and in British intellectual circles, has become the siren song of *NYRB* now that the hostilities have quieted, if not exactly dissolved. In the Frankfurt Book Fair Edition of October 23, 2003, Arthur Schlesinger, Jr., usually a sophisticated and level-headed liberal, signals as much in no uncertain terms in his assault on a "crafty" president who he sees as touched by lunacy: Looking back over the forty years of the Cold War, we can be everlastingly grateful that the loonies on both sides were powerless. In 2003, however, they run the Pentagon, and preventive war—the Bush Doctrine—is now official policy. Sixty years ago the Japanese anticipated the Bush Doctrine in their attack on the U.S. Navy at Pearl Harbor. This was, F.D.R. observed, an exploit that would live in infamy—except not, evidently, when employed by the United States.

The "infamy" is in the analogy. Whatever one may think of the idea of preemption—and I share Schlesinger's negative view of this as a policy framework for the armed forces—to equate the struggle in Iraq with the unannounced, secretive assault on the U.S. Navy at Pearl Harbor simply crosses the line from imaginary history to vindictive ideology. The writing, uncommonly coarse for so sophisticated a scholar; indicates what the expectations of the *NYRB* audience supposedly craves: a Manichean ideology wrapped around purported reviews of books on current events. It would take an intuitive genius to understand what the two significant books by Ivo Daalder, James M. Lindsay, and Fred I. Greenstein are about or what they have to say about President Bush's foreign policies.

Just in case readers of *NYRB* may think the assault on American foreign policy is incidental, it also gives us Anthony Lewis' article on John Ashcroft and Donald Rumsfeld, assaulting them for their roles in American domestic policy—or at least that portion which concerns "the harsh treatment of aliens since September 11." Here at least, the book ostensibly reviewed, David Cole's *Enemy Aliens,* is in accord with the sentiments of the reviewer. Again, the appropriate policy in a post-9/11 environment is an admittedly complex subject, and differences can and should be expressed. It is also an area that is increasingly being reviewed by the judicial and legislative branches, and challenged by the press. But defining the situation as the "Bush administration's attack on civil liberties" and "the Bush administration's abandonment of legal norms" is somewhat broad, since so far only it has affected only several thousand detainees from actual combat zones in which American armed forces were directly threatened. In any event, one would have thought that Lewis's concern for what he calls the "great secret" of America's "commitment to law" is misplaced. It is actually America's great commitment to justice that should be emphasized. And seen in that light, the problem of terrorism might perhaps be viewed more seriously than Lewis cares to admit.

The extent to which *NYRB* views politics and presidents as identical or at least coterminous is highlighted on the other side of the coin: *Bill Clinton: An American Journey,* written by Nigel Hamilton, is reviewed and rebuked in the same issue by Larry McMurtry. Our hero, Bill Clinton, was not toppled by the press (the same people who apparently have fled the field of battle over civil rights and terror) as was Richard Nixon and Gary Hart. Instead, the press was "confounded" by "the brash boy from Hope, Arkansas." McMurtry reminds us that "he kept right at it with Gennifer, kept at it for more than a decade, though with some lengthy interruptions."

But the moral of the story for McMurtry is not the president's steady diet of women, which is what the book is about, but Clinton's surefooted humanism once in office. His kindness to Pamela Digby Churchill Hayward Harriman, wife of Averill, and presumably "the greatest siren of the last century" completes the review. The president as humanist, gives Mrs. Harriman "her one legitimate honor; he made her our ambassador to France. She went to Paris, did her job, and died in the style she had always insisted on, after a swim at the Ritz." With such aristocratic panache, the prequel to the fortieth anniversary issue of *NYRB* comes to a blessed and highly satisfactory conclusion. As with the tabloid style in general, the *NYRB* has readily accommodated its readers to a style of instruction instead of education. In the world of the New Liberalism, one is to read and believe, rather than read and think.

III

If imitation is the highest form of flattery, as the old saying has it, then the *Boston Review* merits consideration for extending the *New York Review* style. But that publication includes serious modifications: starting with a subtitle that eliminates the thorny problem of the word "books." *BR* is frankly "A Political and Literary Forum," and its blunt sans-serif typeface lets you know that the emphasis will be distinctively political rather than literary. It makes a bow to diversity with "New Democracy Forum" then attempts to canvas intellectual opinion on public issues. In the October-November 2003 (Volume 28) for example, the theme is "What Makes Schools Work?" And while the main article and the commentaries share a belief that "vouchers are not the answer," at least some semblance of differences—at least within orthodox opinion—is maintained. It is also the case, that there is far more room given to poetry and poetry reviews than one can find in its erstwhile elder tabloid from the Big Apple. With so few forums available for discourse on contemporary poetry, one must certainly place such emphasis on the credit side.

But when the *Boston Review* gets going on political subjects, it more than gives the *New York Review* a run for its political money. Conspiracy theory is the order of the day. There is Noam Chomsky on "The Bush Administration's Imperial Grand Strategy," which is little more than a restatement of the fashionable view that the War in Iraq is a test case for "the assumption that the United States can gain 'full spectrum dominance' through military programs that dwarf those of any potential coalition and that have useful side effects." The fact that the United States has bent every effort to enlist widespread support, and indeed is clearly ready to weaken its postwar efforts to reconstruct a democratic Iraq, if this will involve other nations, makes little dent on Chomsky. The old chestnut of the military-industrial complex is alive and well in his rhetoric. He knows, for example, "the global wave of hatred" he sees being heaped on the Bush administration "is not a problem." The abstract "they" after all "want to be feared, not loved." Blessedly, this pulp psychology is not couched as a review of books, but simply as a fact to which Chomsky is uniquely privy.

The articles by Juan Cole and Duncan Kennedy are cut from the same cloth. The actual stated purposes of the Iraq War have long since vanished from their discourse. In its place is the raw belief that the United States' main concern is "The Occupation," and that involves longstanding American policies: anti-Communism as a pillar of policy, arbitrary and capricious choice of Iraqi Shiites as the force of choice "to build an Islamic republic." By eliminating the Ba'ath regime we are told, the United States has unleashed the grounds for an extended civil war (between Shiites and Ba'athists) rather than the hoped for civil society. This rather sober assessment is followed by Kennedy's piece which simply asserts that, with the aid of the powerful government and non-government agencies, the Americans are declaring themselves to be de facto colonizers, "the necessary corollary of protecting its multinational investment." Again, these are articles not reviews disguised as articles. And the theme is an unvarying animus. In this piece, the conclusion is offered with a slight hedge, "it seems possible that the second Iraq war will turn out to be the greatest U.S. ethical catastrophe since Vietnam." The line from Nixon, Kissinger, to the two Bushes is made quite explicit—including the fate that awaits the people of Iraq at the hands of extreme Islamists waiting to seize power.

The piece by James K. Galbraith is a further attempt to rehabilitate the late President John F. Kennedy. It turns out that in autumn 1963 he ordered a complete withdrawal from Vietnam. A re-reading of Robert McNamara's 1995 memoir, *In Retrospect,* in which he took personal responsibility for the Vietnam War, bolsters this view. Galbraith's view is that most thinkers from all political spectrums seriously underestimated the extent to which Kennedy gave the order to start withdrawal from Vietnam in the fall of 1963. To be sure, the author is reflecting the strong anti-Vietnam War sentiments of his father, John Kenneth Galbraith. And the idea of a line from the elder Galbraith to Kennedy to terminate the conflict whatever the South Vietnamese government wished, offers a cleaner and nicer filial history. It also places the burden of escalation and contribution of the conflict squarely on the shoulders of Lyndon B. Johnson, who has few friends in Left

Boston-Cambridge circles. But conspiracy theory again saves the day and rounds out the analysis. Johnson may well have shared Kennedy's misgivings. But lacking Kennedy's "determination" he simply capitulated to the military a "nuclear coup d'état." We are informed in the final footnote that the evidence for this is his own 1994 article with Heather Purcell "still available on the website of the *American Prospect.*"

The tabloid form issues into the tabloid style. Any effort at nuance, at analysis in place of attitude, is defeated by the demands of political debate. The terms of that debate are circumscribed by an animus for the president that organizes a left-wing ideology of isolationism. The *Boston Review* 's authors simply never consider that the central core of the American foreign policy is indeed the defeat of the remnants of the Saddam Hussein regime in Iraq. Bush has made it perfectly clear that the fate of the Middle East as a stable area in the world community of nations, and not as the fulcrum of global export of terror, depends on the resolution of the Iraq conflict. The effort to turn Iraq into a present-day Vietnam fails to ask whether such an analogy is realistic, and even less, what the long range consequences of American defeat in Vietnam meant for the current alignment of international forces. The assumptions of the political press or the liberal arts intelligentsia are not open to debate—only its strategic goal: isolationism disguised by United Nations rhetoric.

IV

The *Women's Review of Books,* now in its twenty-first volume year, is published by the Center for Research on Women at Wellesley College. As the title strongly indicates, it has a feminist orientation, while less clever than its politicized counterparts with city slicker names; it has evolved over the years into a thoughtful forum. The October 2003 issue is rich in a variety of themes of concern to women of all persuasions and interests. The most encouraging aspect of all is that the publication is actually a review of books. Even more pleasantly surprising, the reviews are serious in content and subdued in tone.

The near-mandatory review of a new book on Zelda Fitzgerald (this one by Sally Cline) is measured in its appreciation of her aspirations without ignoring the pathologies that this wife of the famous Scott Fitzgerald suffered. Happily, Scott is not blamed for the mental ailments of his wife. An intriguing review of three travelogues by "White ladies" does not shirk the embarrassing role of such works, taking for granted the superior culture of the travelers, along with the unique angle of vision brought everywhere from the Rocky Mountains to West Africa by women. Rebecca Steinitz's summary is entirely apt: "Rather than simply celebrating or condemning them, the best thing to do is read them."

WRB tends to review books that have wide interest in various fields. So a new work by Kathy Davis on *Cosmetic Surgery* is treated with all the anguishing contradictions that the subject calls forth. The reviewer, Gretchen A. Case, points to the "contradictions inherent in cosmetic surgery by considering it to be an 'intervention in identity' rather than a search for ideal beauty." And while identity theory may not resolve the issues involved, at least the reviewer shows the possibilities of good theory in relation to specific problem areas. The same is the case in review of a book on *Black Women Talk about Sexuality* by Tricia Rose. Sharifa Rhodes-Pitts notes that the book assiduously avoids the dominant modes of sexual storytelling, and avoiding placement of narratives into "story containers" such as "rape victim, incest survivor, married women, single mother, lesbian, virgin and prostitute." There is a refreshing sense of concern without strident ridicule.

Throughout there is a sense that reviewers and the journal editors are anxious to grapple with the place of feminism as an ideology embedded within larger issues of race and class. The review by Lori Ginzberg of a new book on *The Syntax of Class* by Amy Lang, points to a broadening of horizons. In addition, the reviewers are not fearful of tackling feminist icons. So in examining Susan Sontag's new book on photography, *Regarding the Pain of Others,* the reviewer, Marilyn Richardson, is quite frank about the concentration of a book on images that lacks images: She intends the pictures she invokes to coalesce inside eyelids closed against remembered images of atrocity. We can review Sontag's take on those we know; we are asked to trust higher analysis of those that are unfamiliar.

The famous and the furious do not get a free pass in *WRB*. That is not to say that the periodical is free from the special interest cant that seems to haunt the tabloid format. One "op-ed" piece asks "Where are the women?" Presumably there is a problem of missing feminists on television—an MIA that certainly has escaped this avid TV viewer. But such lapses are few and far between, and are more than compensated by unexpected pleasures, such as the challenge to the idea that militant women should disdain reviews of cookbooks—and that unabashedly dismisses the older view that what takes place in the kitchen is of central importance to women. Barbara Haber's review essay on the subject is worthwhile in any context.

Having said all this, a question remains whether such special interest publications as *WRB* have a place in the media of the new generation? The publication provides classified advertising soliciting women's applications for a host of jobs in major American universities. This segment, extending from Duke to California, is indicative of the nexus in which these publications operate. They have yet to prove audience viability beyond the confines of liberal arts departments in major universities. They also serve to drain off the participation of quite able younger scholars from broader based publications of a viable sort. Premature obsolescence seems to haunt such special interest publications that care about at a time of lower levels of mainstream acceptance. What happens when that takes place, when such women become the driving academic force, is an issue likely to be felt and faced in short order—when the nature of professional life itself comes under greater scrutiny.

V

Ruminator Review derives its title from the grand tradition of "ruminations." The sub-title is *The Independent Book Magazine*. But if we are to believe its editor, Margaret Todd Maitland, independent ruminations have well defined ideological limits. We are told in her editorial that there is no perfect family. "Though I'm offended by the rigid definition offered by the religious right in this country—and outraged by legal attempts to limit who can and cannot be considered a family," our editor does concede that "it's not easy to dismiss the archetype"—presumably this flexibility was learned from observing a family of elk in a visit to Oregon by her own family. Published at Macalester College in Minnesota, issue number 15 (Fall 2003) has taken the tabloid quarterly.

This relatively recent entry into the field has an emphasis on environment, with each issue having a thematic—in this case, the family. The previous issue was on rural America. The books selected for review tend to be off the beaten path and the reviewers still more so. In a lead review on adoption and kinship, the reviewer, Cheri Register, rebukes the author of a book called *The Wailing Child* for its hyper-extended usage of adjectival writing, and for good measure at her "Christian imperialism" for daring to raise "the message of faith." In the same vein, in a review of an edited book on *Gay Men Write about Their Fathers,* the reviewer, John Townsend, commends one essay in particular for its "most blistering, not to mention politically incorrect" essay on a father as well as a son who are both "gay." Given the horrors of the narrative extract, gay might not be the appropriate word. The two men never reconcile, and the essay is cited for its "courageous honesty" that father and son "never got to mellow."

Perhaps the highlight, or "lowlight," of the issue is by Jacqueline White, self-described as a Twin Cities writer and director of "Project Off Streets and the Gay, Lesbian, Bisexual, and Transgender Host Home program." At least she cannot be accused of having a narrow vision of alternative families. She concludes her essay (again, the op-ed phenomena at work in terse tabloid formats) thusly: "[W]hen I was dropping off a young woman I could barely make any sort of claim to at all, when she was jogging across the street holding down her fedora against the wind, that I understood just how transient the whole enterprise of motherhood is. Parenting is evidently one long leave-taking, and our affection must negotiate that ever widening expanse."

Interview, essays, opinions, and even reviews on all sorts of alternative family systems follow. The quality of writing strains for difference but the total package ends in sameness. The issue concludes with a proclamation attacking Section 215 of the Patriot Act which raises the potential of the FBI obtaining records of those who make book and allied purchases in the search for enemy agents or terrorists. The First Amendment is indeed unduly abridged by such legislative authorizations. Then again, the source of such concerns *Ruminator Review,* would be more credible if it practiced what it preaches, free speech. Just about the only enemy in this particular issue is the so-called conventional family. The disfigured people found in the books herein reviewed would indicate that perhaps traditional values offer more to keep people "off streets" than do newer forms of social work and self-help among the deviant crowd. If one is to believe the parade of reviewers and commentators at Macalester College, the family according this publication is less a work in progress than in disintegration. Still, this publication is indicative of the ability to do much with little—no small asset of the tabloid format.

VI

It requires reiteration that the generalizations drawn from these various publications are based on special and perhaps non-representative issues. Generalizations are treacherous in the best of circumstances. But a failure to grasp the implications of new styles of academic work on the ideological battleground carries even more serious ramifications. So while I am willing to concede that later events and issues may disconfirm some of my remarks, for now I am convinced that serious changes are underway in the life of the mind.

To start with, these new formats are desperate attempts to break out of the isolation that many in academic life feel whatever their ideological persuasion. The *Claremont Review* published in California is a strong effort to emulate the success on the right of the *New York Review* on the left. It too seeks to frame the books that it reviews, and by whom, into a frontal assault on the dominant liberal cultural establishment. A format is not owned by one particular belief system, although it is a curious fact that the hard left has made much greater use of the new form than any other segment of university life. But the fact is that these publications, whether edited off or on campuses, appeal directly, if not exclusively to university personnel, and to those interested in liberal arts concerns.

That in itself becomes a conundrum of sorts: for the format that harkens back to mass appeal, can barely get beyond high-class appeal. And with the arguable exception of the *New York Review,* which boasts a subscriber base hovering at the 100,000 mark, none of these newer publications have broken through to a larger market. The near exclusive advertisements by university presses in these publications are itself a tell-tale sign of the limits more than the goals of this market.

A second problem is the issue of cachet. Does publication in these tabloid publications carry with it the potential for personal and or professional advancement? The quarterly journals may serve even smaller audiences, but they are recognized as the source of scholarship. They entail editorial boards of peers, review processes that have been maintained over time, and a style of citation and referencing that is recognized as a guide to keeping out screed and scare, no less than including sobriety and rationality. The steady drumbeat of political issues, especially those at the most general levels of news stories and presidential decisions, may be momentarily engrossing, but whether they do more than orient the bulk of liberal arts personnel to what constitutes politically correct thought at any given point in time is difficult to say.

A third area, and perhaps the most decisive, is the taken-for-granted characteristics of so many of these tabloid stories and reviews—whether a broad readership exists among the intelligentsia for such ideological force-feeding remains on the table. While it must be acknowledged that these publications do have a heavy impact on the political culture of the academy, especially the advanced schools in which a certain orthodoxy is a welcome relief from real thought, doubt remains that such partisanship can actually muster a quorum at the end of the day. These publications breed their own core group. The same names appear and reappear. They become captive to a cadre of writers, and hence tend to rule out of participation even of those acolytes prepared to go along for the ideological ride. It is by no means certain that the authority vested in scholarship translates easily to the sound and fury of these tabloid publications. After all is said and done, the tabloid newspaper is designed to be read quickly, and discarded by the next day. This is not exactly the prospect that academic pundits have in mind that prefers to think that they are writing for all eternity.

If I had to make a wager on this new form of information delivery, the academic tabloid, I would bet on some publication like the *Chronicle Review,* which comes as part of the *Chronicle of Higher Education.* Like its ancestor, the *Times Literary Supplement,* it is attached to a major publication that seeks and receives wide public support amongst relevant audiences. Embedded in some larger program, however independent in contents, such supplements build upon a longstanding and honorable tradition of cultural ideas and opinions. It serves as a resource base that make possible informed approaches without a corresponding highly refined ideological agenda. What seriously curbs so many of the new tabloids is not the strength of its convictions so much as the absence of surprise in their presentation.

Whatever the long-range prognosis for such publications, it is evident that they have left a mark on the American culture in general, and publishing in Anglo-American in particular. In the struggle for a broader base is a demand to be heard beyond the boundaries of the academy. But it is so only if the search for a broader audience, one that embraces everyone from the *Atlantic Monthly* to *Science,* can actually produce a cultural advancing over the present bifurcation of mass and elite. That is unlikely because these older and estimable monthlies cater to an undergraduate audience of years past. They have yet to define a serious role in a world of the academy and policy as it now stands. More probable is a growing disparity in the cultural arena between the ideological and the scientific. This may not impinge on the political winds that dominate American life, but it should insulate, perhaps even isolate the scientific community from shocks of the moment. If it serves only to pan—at the political machinery or the cultural apparatus, if it can only celebrate the latest candidate to put himself or herself forward in the national limelight, or champion any form of deviance with a following, then tabloid scholarship will be to scholarship what martial music is to music: a sad reflection on the current plight of intellectual life in American society.

The problem is that the cultural apparatus has invaded and infested the scientific community: everything from the debate on global warming and environmental safeguards to the shift in sociology from stratification analysis as a scientific discourse to race, class and gender mantras of anti-American nastiness are at work. Coupled with the uses of religion and theology to advanced special social and political causes, it is hard to imagine the cultural superstructure being able to back away from its respective missions. The one saving grace is public opinion itself. The cultural icons of yesterday have far less cachet now than forty or fifty years ago. For better or worse, new images push aside old ways of thinking and responding to events. Beyond that is the sharp reduction in the outreach of university presses and professional organizations to impact much less control public opinion. Political publications, however aggressive, are responsive to fiscal constraints. And when the tired alibis that these are a result of recessions fade into the night—as they should—they will all be left with the need of responding to the will of the very people whom they hold in such sad but obvious contempt.

11

Scholarly Pornography

In January 2005, one of the premier scholarly publishers in the English language, Princeton University Press, published a brief, eighty-page pamphlet in book form, called *On Bullshit* by a well-respected philosopher, Harry G. Frankfurt, who had written widely on basic themes in epistemology. The titles of his previous works indicate the subject-matter: *The Reasons of Love*, *The Importance of What We Care About,* and *On Truth.* The concerns addressed range far and wide in the history of philosophy, from Augustine to Wittgenstein, and do so with intelligence and appreciation.

The book in question, *On Bullshit,* received a wide range of critical responses in reputable publications, ranging from "defining the essence of postindustrial society" (Scott McLemee), and "the humor and the naughtiness lie in the contrast between the highfaltin' and the indelicate" (Roger Kimball). For one critic the effort was a metaphor for the Presidency of George W. Bush: "We are drowning in bullshit. I mean the Bush administration has practically made it a Cabinet position" (Dan Neill). It would take a strong will to make the counter claim that we are not wallowing in humbug, or that our life and times are exempt from the history of hubris. After 175 reviews of this monograph, the inevitable happened. One year later, in 2006, Graham Edwards produced *The Business of Bullshit* and Nick Webb offered up *The Dictionary of Bullshit*. Alas, although Professor Frankfurt's extended essay is a proper attempt to distinguish academic practice from professional principle, it ends up discarding the latter by casting a wide net of moral suspicion on the current status of learning as a whole.

What brought back such recollections of this recent success in university press publishing is less Professor Frankfurt's monograph, than the inevitable effort at imitation, not so much of the contents of that volume as the shock value of obscene language as a measure of the courage of the cowardly. We now have a volume of philosophical discourse for the learned class entitled *Mindfucking* which attempts to extend the boundaries of epistemology, or at least to demonstrate that the field is not the preserve of ancients and fossils, but can be practiced with ease by campus-dwelling freshmen.

This work, by another professor of philosophy, Colin McGinn at the University of Miami, was issued in Canada by McGill-Queens University Press, a distinguished university press in its own right, one distributed in the United States by an equally venerable house, Cornell University Press. It seeks to investigate and clarify "modern techniques of thought control. Professor McGinn assembles the conceptual components of this most complex of concepts: trust, deception, emotion, manipulation, false belief, and vulnerability." And as a display of one-upmanship, philosophy itself is seen as "a type of mindfuck. A vast litany of psychological characteristics from jealousy, disorientation, insecurity, and prejudices, are adduced as an aid in this mindfuck. The result is delusion and even insanity." So this can be viewed as a protection against insanity. Again, in seventy-six pages, with the aid of a keen apparatus drawn from the history of philosophy, we are urged to use expletives as therapy.

The promotional effort for this work offers a plain guide to the perplexed: "Being surrounded by bullshit is one thing. Having your mind fucked is quite another. The former is irritating, but the latter is violating and intrusive, unless you give your consent. If someone manipulates your thoughts and emotions, messing with your head, you naturally feel resentment: he or she has distorted your perceptions, disturbed your feelings, maybe even usurped your self. Mindfucking is a prevalent aspect of contemporary culture and the agent can range from an individual to a whole state, from personal mind games to wholesale propaganda." Just where the line is to be drawn between education and manipulation remains the mystery in this appeal to natural goodness and perfection. But this is a secondary concern that Professor McGinn can work out more directly with Professor Frankfurt.

The use of coarse language introduces serious issues in ethical theory, which is assuredly a strong part of the history of philosophy. It also expands the limits of taste for presumably the most elevated segment of publishing. Such large questions merit attention: Does criticism of language condemn the critic to being an old fogey, or worse, an old fart? Does the use of such expletives add to our knowledge of the subject matter of the field, or aid in studying Kant or Wittgenstein? Are we better able to avoid "mental manipulation" by employing such expletives? Or put in reverse order, does the removal or censorship of such "curse words" detract in any way from the subject matter at hand? Finally, do those university presses publishing such works extend such tough street rhetoric to include criticism of all the popular myths and ideologies of the time, or is a line drawn ruling issues of race, gender or class off limits? More pointedly, does the new freedom in expletives open up new vistas of analysis or does it effectively seal off such areas in favor of cheap linguistic thrills? It comes to pass that large issues of liberty and license are very much part of drawing the line either between or after bullshit and mindfucking.

I will leave to the professional philosopher responses to these multifaceted queries, and will simply address the issue of the appropriateness of these volumes for university press publishers. If these books are part of the discourse on basic issues in the history and status of philosophy, will the language used in the titles promote the larger concerns of civil discourse? If this unique community of roughly one hundred publishers in North America fails to address this question, then it seems perfectly within the bounds of good judgment to question whether the university press will stand for anything distinct from any other segment of the publishing community. Beyond that, what is the connection between the university press and the university community as a whole? Is the latter simply an extension of the social nexus as a whole, without special norms or responsibilities?

More pointedly, if university presses in their perfectly reasonable desire to survive and expand their sphere of influence, both within the higher learning and in the mass society as such, must resort to pornographic language, pure and simple, to sell their wares, what distinguishes them from commercial life as such? It is hardly a secret that the margins of profitability in university press publishing are so narrow as to often require subventions from general university funds. Even those 10 percent of university presses that are relatively sound in fiscal terms are expected to contribute to the general welfare of the universities in which they co-exist—or at least not draw from general funds.

The problem is not only implications of efforts to sell to a market through the use of shocking terms, but also the potential impact upon core curricula. Will professors be advised to juice up their course offerings? Are we to suggest that instead of calling a course topic sexual behavior we substitute the terms fucking interaction? More pointedly, is the classroom to become a vehicle for self-examination or just plain exhibitionism and experimentation? By the same token, if a student is to find himself or herself in disagreement with the professor, is the proper comment: "Sir, you are a bullshitter" (to the roars of the fellow students) or "Are you a mindfucker?" Does such language become the end all and be all of discursive behavior among students and teachers, or for that matter, teachers and administrators?

I suspect that the uses of such rhetoric, far from having a liberating impact, in fact serve to curb dissent and discussion of serious issues. Those who carry weapons of grading or sheer physical intimidation, hardly an unknown property in college and university life, will be granted a license to employ precisely the instrument of emotivism in general and cursing in particular as a means to quiet dissent or constrain opposition. The use of such rhetoric forecloses discussion, and rarely opens new avenues of thought. The history of thought, whether of radical, liberal, or conservative ideas, is conducted in a common linguistic discourse and civil conduct precisely in order to bridge the gap between ideas, and to reach out to people in non-intrusive, non-menacing ways.

The problem in the *metaphysical* assumption that profanity is a source of liberation inheres in the *logical* equation of language and the higher learning. There is not a shred of evidence to indicate that the unrestrained use of taboo words (rightly or wrongly imposed by the larger society on the individual at a particular moment in historical time) is an indicator of a superior person in intellect or even less, a person capable of ethical conduct conducive to a complex society. Indeed, I am not even sure that such a claim could even be made for simple societies.

The extent to which the uninhibited use of certain terms is an advantage or a disadvantage to a community or a society is determined by its explanatory powers, its ability to resolve empirical or ethical issues that one must cope with on a daily basis. It might well be argued that the use of terms like "bullshit" and "mindfucking" we so widely employed to cover a range of emotions and preferences that this search for meaning, for philosophy itself, is deadened; frequently permanently so. One derives the meaning of those who use such language by ignoring the curse words and seeking answers in the remaining part of the sentence or the emotive contents of the comment.

The emotive impact of terms like "bullshit" or "mindfuck" has the power of weapons behind them, not the weapon of power as both Professors Frankfurt and McGinn doubtless intended. I submit that behind the courage of low-level quotidian slang is high-level intellectual reticence, incapacity or at least unwillingness to come to terms with fundamental issues of personal psychology and public philosophy in ways that promote the advancement of civilization as such. University presses are under competing and contradictory pressures: they are set up as paragons of quality (if not of virtue), but then demands are placed upon them to show a surplus, or more pointedly, provide a profit to the university at large. In a sense, this is a new phase in the struggle first observed by Charles Homer Haskins in *The Rise of Universities* between the medieval inheritance and the commercial realities of modern university life.

What is particularly questionable is what, if any, response can be offered to name-calling of this sort for the purpose of distinguishing bombastic carrying on and the place of legitimate authority. This is especially the case when it comes to universities that derive their very sustenance, commercial as well as intellectual, by presuming a difference between those who know and teach and those who do not know and learn. Appeals to evidence, experience, and empirical information are grounded in more than a primitive positivist creed that to state a preference is simply to put forth a bias. Wittgenstein has often (and wrongly) been charged with offering a solipsist agenda for the denial of a real world apart from the language that expresses its contours. These jeremiads offered by Frankfurt and McGinn only serve to rekindle a mid-twentieth-century argument that analytic philosophy and scientific method have happily moved beyond.

University presses that promote such titles in search of meaning would be well advised to take seriously the risk factor in becoming successful commercial publishing agents. They may actually wake up one day to find themselves having found fiscal success in the process of losing the aim and intent of scholarly publishing as such. When the shock value of these titles wear off, through excessive usage if not in revulsion, and that will be soon enough, the world of the higher learning will be left to wander about in the lower depths—and with no light in sight. As the non-academic underworld often declares: money talks, nobody walks.

References

Colin McGinn, *Mindfucking: A Critique of Mental Manipulation* (Montreal: McGill-Queen's University Press, 2008), 82 pp. + xiv.

Harry G. Frankfurt, *On Bullshit* (Princeton, NJ: Princeton University Press, 2005), 80 pp.

12

Publishing Programs and Political Dilemmas

Publishers do not wring their hands over ethical concerns. Indeed, adherence to norms of behavior and conduct are taken for granted in every aspect of scholarly publishing. Without such assumptions, it would simply be impossible to make decisions and act. Our own Transaction list is now 5,000 books deep—or old—as the case may be; this was achieved without getting lost in ethical quandaries.

Ethical issues do become important when the norms of behavior and the ordinary rules of conduct that "everyone knows" are violated. Often this occurs for substantial reasons. Very rarely are ethical matters invoked in relation to monetary outcomes. There simply is not enough money to make conscious or unconscious violations of norms of publishing worthwhile.

What we do find is that ethical norms are shared for "good" reasons: to improve the image of minorities, to cast in a harsher light than is warranted the evil of one's enemies, or simply to give added weight to an old theorem that in fact would be badly shaken by a plain statement of the facts. Of course, such norms are called into doubt when self-serving interests are at stake.

At Transaction, ethical issues are quotidian confrontations. It would be unfair to cite specific titles or authors. But in 5,000 titles there are obvious cases in which personal bias, professional pique, or ideological proclivities preempt the norms of scholarship in social science. Indeed, the very "softness" of so many areas in which Transaction publishes heightens problems that broadly speaking take an ethical twist.

That said let us move ahead to the specific issues. Hopefully, the approach taken will be of some value to those who contemplate a life in scholarly publishing. For the hand-wringers, for those to whom every issue from bathing to baseball is laden with moral drama, this sort of "neo-positivist" response to ethical concerns in publishing will simply be unacceptable. These people are, of course, entitled to stop reading immediately and go directly to a higher authority: either of a providential or presidential sort.

Multiple Authorship: Multiple authorship is less a matter of ethical dilemmas than egotistical preemptions. Authors who work together tend to maximize their own involvement and inputs, and as a consequence, minimize the work of colleagues. I do not believe that I have ever met a participant in a co- or multiple-authored work who ever acknowledged that a colleague might have done more than he or she.

This may play out in contractual problems: namely the relationship of percentages in royalties between co-authors. We had one bitterly fought out case, in which one author of a basic text in methods of research claimed to have done 60 percent of the work and therefore wanted an exact recompense of 60 percent of the royalties. The other author felt that to enshrine such a ratio would be a *de facto* as well as *de jure* expression of his worth. The difference in dollar amounts would hardly buy a meal for two at Burger King.

The resolution in this instance was to permit the 60-40 ratio to stand, but to provide the 40 percent author with first place on the title page and promotional information for the book. This worked well in this instance—it may not in others. Not to say that every confrontation of co-authors can be handled with Solomon-like aplomb, but the conversion of the ethical into the operational can save a great deal of wear and tear on publishers and authors alike. Given the rising tide of multiple authorship, publisher-generated texts, and agent involvement in the creative process, such practical wisdom may help.

Sometimes a co-authorship is wired to a mentor. This can lead to the sort of complex dilemma like the Imanishi-Kari-Baltimore matter. In this case a young scholar (Dr. Thereza Imanishi-Kari) listed a distinguished researcher and former president of the Rockefeller University (Dr. David Baltimore), as co-author of a project later assaulted by another young scientist Dr. Margot O'Toole. While in this instance both researcher and senior scholar were exonerated after a series of appeals, some of the attention might have been diminished by calling a moratorium on the awful practice of listing the senior professor or laboratory director as co-author, simply because he or she was a sponsor or provider of the facilities.

But while problems like this clearly involve ethical issues of substantial dimension, they cannot be resolved by the publisher as such. To attempt to do so would be, to put it mildly, presumptuous and rude. Publishers of scholarly materials must rely on what Blanche calls, in *A Streetcar Named Desire,* "the kindness of strangers." For authors are for the most part strangers, as are those who review their work, and even a university letterhead cannot be viewed as a sworn testimony of moral probity.

Peer Review Process: There are perversions and distortions in the review process. Referees make all sorts of judgments that are biased, wrong-headed, ideologically motivated, and just plain wrong. But as Winston Churchill noted about democracy as such, it may be an imperfect system, but can anyone offer a better one for people to live under?

Peer review admittedly works with less than perfect resonance. But if the issue is willful exaggeration of worth or, as is more often the case, willful denial of worth, then I must say that little of this goes on. Of the more than fifty serials which we publish at Transaction, I can recollect only a handful of cases in which anything resembling solid evidence of ethical misconduct arose—and this over a forty-five-year time span. Of course, authors spurned are authors angered. But that is a far cry from proof of bias.

The safeguards that exist are considerable and operational: (a) the norm of confidentiality; (b) the norm of multiple reports, especially in complex manuscripts; and (c) the norm of external solicitations, that is, from scholars not directly related to the campus or person of the manuscript being reviewed. To be sure, limitations in the expertise process limit perfect ethical conduct. Ethical breakdowns are predicated on real or imaginary differences between experts in the same field on a select topic. This I take to be essentially a non-ethical problem.

We also need to recognize that peer review is usually for one publishing house or one editorial product. And in the United States, where there are, to put matters mildly, a plethora of journals in just about every field of intellectual endeavor, even if there is a clear-cut case of ethical bias, the prospects for publication in an alternative publication remain relatively high. What is amazing is not how little scholarly material is published but how much.

Again, this is not to be dismissive of problems of difference in relative standings of journals. Nonetheless, it is also the case that high status accrues to journals because of the high quality of its published materials. This would indicate that violations of ethical norms in the review process are necessarily limited precisely by the quality of the journals involved. I would even formulate this in axiomatic fashion: the higher the status of a journal, the lower the number of instances in which ethical norms are violated, or even claimed to be violated.

Editorial Networks and Quirks: I confess to be somewhat un-sure of why this is posited as a major ethical consideration. I suspect that the question is to what extent publishers establish networks, either formal or informal, and hence limit certain people from access to the print media; and the extent to which this networking is not so much based on professional consider-ations as politically correct consecrations. Having interpreted the issue, in I hope not too improper a fashion, let me try to pro-vide some answers.

Yes, of course, there are discussions and interactions between and among publishers and editors. Yes, of course, there are net-works that determine who is acceptable and what is preferable. That some of these interactions may have "quirky" dimensions is also clear; although one might say that eccentricity and discov-ery are not exactly strangers to each other, nor should they be. Having granted this premise, one must be careful not to presume conspiratorial modalities amongst publishers. Differences between publishers, especially in scholarly communication, are as much a function of the market as it is of morality.

Different publishers come to be known for signature elements in publishing; thus one finds many smaller presses filling or occupying certain niches in the market—and these can be from feminism to regionalism, no less than from poetry to science. No single publisher can fill all the needs of the marketplace. And in a democratic order, the choice, the possibility of alternative vehicles of publication is fundamental. The need to survive in the marketplace is a limiting condition on quirkiness or eccentricity, perhaps too much so! Likewise, collusion among publishers is far less a problem than their seeming inability to rationalize even elementary issues of distribution, that is, getting their books to the market. Other sectors do a much better job at industry-wide association and information sharing.

A far more serious consideration than ethical matters, from my viewpoint, and one that threatens the very nature of choice in publishing, is the continuing pattern of monopolization. What one finds is that a chimera of health exists in the form of a seemingly ever-expanding number of small publishers, but they have a dwindling market share. The inexorable logic of big publishing, linking itself to big chain store retailing, and both linking up with major media servicing in other areas such as television, World Wide Web, and newspaper publishing, for example, is a far more serious threat to a free and unimpeded exchange of ideas than the virtually nonexistent collusion or cooperation of publishers.

The question of networks, really the question of collusion, strikes me as the sort of rather minuscule issue raised by a disgruntled or unhappy author who feels that his or her manuscript, having been seen and rejected by eighteen publishers is really a brilliant breakthrough, and that the publishers have all managed to collude to prevent publication. To such an author I would suggest sending the manuscript to a nineteenth publisher, or more realistically, sitting back and reviewing the work to see what is wrong with it.

Now there are cases in which many publishers reject manuscripts that might merit (with substantial revisions) the light of print. Every publisher has books on their list of which they are neither proud nor pleased. And after 5,000 books published, Transaction has had a few of these as well. But these are not the core of our list. That we published them is not an ethical issue because we had no moral qualms. Rather they represent a special judgment that a certain book, however eccentric to most, is worthwhile. An ugly duckling to the multitude can be a beautiful swan to the unique publisher. But having said this is not to raise the prospect that most publishers behave unethically. It is simply that one publisher saw possibilities in the realm of either ideas or sales or both for a particular manuscript.

Author-Publisher Misconduct: Here we come to a realm in which the relation of ethics to law is central. In the older Latin and Germanic language traditions (*recht, derercho,* and *droit*), legal and ethical concerns are bundled into a common core of discourse, making them easier to deal with. After all is said and done, terms like misconduct and dishonest have rather precise legal meaning—and everyone connected with the information environment needs to be familiarized with such legal norms. Where then do ethical judgments enter? And here I would suggest that they do so when huge domain assumptions are at war with each other.

For example, the right to know as a struggle with the rights of property is one such area that still angers the publishing community. But it does so not as a purely ethical issue, fought out between Kantian and Hegelian titans, but between publishers and librarians who may not even know of such philosophical antecedents. Norms are set within legal frameworks established by a legislative branch, pushed forward by an executive mission, and interpreted by a judicial body. There is no perfect resolution to such big issues. Domain assumptions are hacked away at rather than upheld or overthrown. Depending upon a specific decision from a specific court, publishers become joyful while the librarians become sullen (or vice versa). Ethical mandates in such a context lose much of their emotive punch.

So what we are looking at are cases of author misconduct, that is, from fraud to multiple submissions, that is the stuff of everyday publishing life; and also publisher dishonesty, failure to pay proper royalties or unfairly making charge backs to authors for alterations or costs that never should have been passed along.

These transgressions do of course occur. Sometimes they are innocent mistakes; at other times they are a calculated attempt at deception. They are widely reported, and serve to discredit and inhibit such conduct by other publishers. It serves little point to emphasize such imperfections and indeed, draws attention away from the larger fact that most publishers and most authors operate with integrity. This makes possible the community of ideas to begin with. Indeed, in the current technological situation—in which inventory records are linked to warehousing and sales records, royalties revenues simply become one aspect of the accounting system. It would be more complex to distort royalty regimens than it was say fifty years ago, when such auditing entries tended to be manual in character.

There are of course publishers who operate in the gray area sometimes referred to as vanity publishing. These publishers promise performances that they cannot and have no intention to deliver—especially in areas of marketing and production. It is easy to look down one's nose at vanity publishing in scholarly areas, but in fact, most vanity houses go after non-scholarly material and people, knowing well that authors appreciate the extent to which the status of the publisher is a highly determining element in the receipt of the work itself. Dismissing vanity publishers is easy enough, and requires no great moral outrage to do so. More critical, and closer to the home of all scholarly publishers, are subventions to produce works that otherwise would not get published.

There are a myriad of such subventions: ranging from the university wish to support a press program however poor in output or performance (often for the purpose of satisfying a local academic constituency wishing to be published), to major scholarly houses that accept grants, nay insist on them, in order to publish large, omnibus collections. And here we need to simply recognize that foundations, institutions with special interests, no less than universities, are constantly suggesting prospects to publishers in which decision making is fudged so that the quality of the manuscript and the amount of the underwrite get fused in the decision-making channels.

These are not easy considerations. Often involved are works of major figures—from Edison to Jung—and support from the

Urban League to the RAND Corporation—again, quite worthy
and honorable agencies—for works that might otherwise not be
published for want of a large market, or presumption of such a
lack in market potential. These are very difficult decisions for any
scholarly publisher, and they arise with increasing frequency as
the economics of publishing comes smack up against the necessity to publish solid materials. At times, it may well be the case
that a publisher accepts a work or a series of works that it would
otherwise not publish, and perhaps ought not be published.

Rationalizations for such publications vary. Some say subventions make possible expenditure of funds for other works lacking
support. These concerns are similar to other college and university
decisions, for example, with respect to tuition. A full tuition is paid
by a mediocre student, which then permits the institution to bring
on board a brilliant student of limited fiscal means. This suggests
the extraordinary complexity of decisions involving judgments
of ethical merit in decisions about economic worth. Perhaps the
scholarly press should be held to a higher standard than the university community in general since the norms of excellence that
operate in the former are more intense then the norms of participation that characterize institutions of higher learning as such. But
in such matters, a priorism helps very little. But having said this,
we must not disguise the existence of what may be construed as
vanity decisions in supposedly pure scholarly publishers.

There are elements over which a publisher simply has no control.
It is impossible, no matter how tight the peer review system may
be, to detect every cause of fraud or misappropriation of the work
by one scholar of the production of another. This is especially true
in contexts of multiple authorship, and in areas of such exactitude
as to prohibit even the best of scientific publishers to discover the
misdeed. One might claim that ignorance of the manuscript is no
less inadmissible than ignorance of the law in passing judgment
or sentence; but we have to be thankful that such claims are rare
in the life of a publisher.

The decisive ethical problem often enters *after* such discovery of
wrongdoing has been made. Does the publisher remove the book
or the manuscript from circulation? Ask for extensive revisions?

Determine the transgression to be of such slight consequence that it can be ignored with impunity? Then there are cases when after a work appears criticisms appear to indicate that the author has racial or religious biases of unacceptable dimensions. Does the publisher remove the book, quietly let the book perish of marketing attrition, or ignore the critics and reassert the right of the author to be heard despite criticisms, in the presumption that bias in public opinion may be greater and more damaging than in a particular author or title?

I confess not to know how one answers such matters with a categorical imperative. Contextual considerations no less than manuscript content enters into the decision-making process. Publishers respond to similar manuscripts differently. There is no way to adjudicate such items. What we have given the plethora of publishers and multitude of authors is a very large set of choices about which a finite number of decisions is ever possible. The democratic nature of the publishing environment is the best assurance against abuse. The normative nature of the academic environment is the best assurance against bias. In an imperfect world, this is the best way to treat issues of moral concern. We search for the ideal and live in the real. We aim for the better, but must always deal with the lesser. On the other hand, when ethical judgment is viewed as a sensitizing element in decision-making, we all benefit: publisher, author, reader—and this holds whatever the format—printed books or electronic transmittal.

The ethics of publishing, as ethics in general, takes place in contexts of the possible. Those who wish to dwell in exclusive realms of categorical imperatives may have splendid careers in the ministry, but they are going to have a hard time in publishing—and that holds even for the marvelous world of scholarly publishing to which I have dedicated a lifetime of service.

What one must confront in decision-making about manuscripts, topics, authors, and personnel is rarely a choice between good and evil. Rather it is a choice; a selection is a better term, between goods. Very few people in the realm of scholarly publishing perceive themselves, or can be perceived by others as behaving unethically. As a result, the agony of decisions revolves around close calls: whether to publish a very good manuscript that one knows full well has virtually no market support; whether to publish a manuscript that is unpleasant or ideologically removed from the normal biases of the age; whether to place resources—always scarce—at the disposal of personnel who merit higher salaries or scholars who merit wider marketing effort.

The quotidian life of publishing does involve ethical issues. But they are often wrapped and bundled in disputes between authors with each other, between authors and their editors, between editors and each other. Then there are a range of allocation considerations with strong ethical dimensions: how much for "content" and how much for "show." Does one buy an advertisement for a book or orthopedic chairs for customer service employees? To speak in such terms is not to make trivial matters of ethical judgment. It is, I noted earlier, an appreciation of contextual concerns in making such judgments.

I was trained in the great tradition of naturalistic ethics. And while I have come to appreciate deeply the importance of religious commitment and resolve in the conduct of human life, that has in no way lessened my appreciation of the context of content, or as it was known earlier in the century, the experiential grounds for confronting the world. There is always a risk in debasing ethical judgment by converting every empirical proposition to a hand-wringing moral concern. To do so is to lose any distinction between the weighty and the flighty.

Ethical considerations in publishing are real enough, without confusing them with ordinary decisions and events. The need to publish is a correlate of the need to learn. This is hardly an ethical matter, but it goes to our essence as human beings who communicate with each other through language. What we publish, how we publish, when we publish, and where we publish, does involve ethical considerations. The capacity to look at ethics in an empirical manner is far more serviceable than the tendency of absolutists to convert every empirical consideration into a moral imperative. I can only say that for me such elementary considerations make possible a decent publishing environment—one that can take place without bombast or hokum, but with a genuine appreciation for limitation in all human beings, and yet, with the Kantian necessity to strive to overcome such limits.

13

Political Periodicals in Policy Formation

Walter Lippmann was fond of saying that "The news and the truth are not the same thing." This is at the core of political periodicals—of somehow squaring this dichotomy. It is a curious fact of American letters that between the technical journals written and read by professionals of varying scientific stripes, and the general newspapers and commercial magazines supported essentially by a combination of mass readership and even more massive advertising, that a cluster of periodicals in print form and online that have limited circulation, manage to wield enormous impact on the policy bodies in both government and the private sector. Personal influence can and does translate into political influence. The existence of "mass society" blurs but does not destroy the power of elites; those powers are serviced by specialist publications and information services that become more rather than less important in the present era.

Nor is this element in the political process limited to the United States. Information derived by "bloggers" in nations such as Cuba, China, and Iran help explain crisis events, and no less, limit the authoritarian tendencies of the regimes in such countries. The Platonic tendencies toward stratification extend to information no less than the exercise of formal authority. In that sense, publications with overtly political agendas remain very much part of the publishing landscape. The fears of Orwellian prophets exist—the massive power of the State to regulate information flows. But then again, the converse is also true; the massive amounts of information limit authoritarian tendencies in ways unheard of and unparalleled in past centuries.

"The Politics of Publishing," served as an organizing premise and launching pad to determine how information providers and ideological constructors engage in serious analysis of the role of this sort of hybrid publication in American life and letters. I say hybrid since, in terms of personnel, they often involve a cross between journalists and scholars and, in terms of audience, they often appeal to professional scientists and academics who have a heightened sense of public awareness, but not necessarily a desire much less capacity to translate that awareness into more active forms of political behavior.

The weeklies and monthlies represented in this convocation included *Commonweal, U.S. News & World Report,* the *Nation,* the *New Republic,* and the *American Spectator.* The monthlies that were represented included such major opinion leaders as the *Washington Monthly, First Things,* and the *Wilson Quarterly.* Among the quarterlies represented were *Partisan Review, Freedom Review,* and the *Independent Quarterly.* I was able to confirm some premises with which I came to the conference and disconfirm others, while still others remain to be resolved at a later date. For the explosion of information in electronic formats change dramatically the range of political considerations, and no less, the outreach of specialist organs of political opinion.

My interest in this topic is somewhat akin to, but also quite distinct from, the work of Charles Kadushin, who examined political publications for their reputation and public impact on larger policy concerns of the nation. My own concerns are what might be termed the softer general forms of knowledge and policy, namely information and ideology. This focus coincides with my own long-standing interest in the sociology of knowledge. In addition, my interests are in the ways in which policy formations are generated by publications—whether in hard copy or electronic forms, and if they employ social scientific rather than strictly political party or ideological talent. In this, perhaps the perfect model was the *Public Interest,* which was solidly rooted in social science personnel at the editorial and authorial levels alike, but yet clearly served highly focused policy interests that were not uniformly endorsed by the body politic. The fact that such a powerful voice in the policy arena was unable or perhaps unwilling to survive the heat of battle itself tells us that the impact of such publications is significant but finite. No effort has been made to gauge the specific impact or reputational standing of each publication either directly in the marketplace or with respect to a rank ordering. The intent to evaluate political publications and blogging was not to establish a hierarchy, but a personal summary and estimate of the political ideology of policy organs.

Perhaps the most elementary, yet easily overlooked aspect of the editorial direction of each of these special publications is the raw intelligence and dedication of the editorial cadre. Whatever their point of view on specific issues, they share in common a sense of the importance of quality in the dispersal of information, a belief in the enlightenment canon that solidly grounded views will prove persuasive in shaping and changing the minds of people, and a broad-ranging civility that derives from a respect for the printed word over and above partisan ideological visions. While such kind words may not extend to information on the Internet, the belief that words count and ideas matter are shared by all varieties of political opinion makers.

The explicitly political periodical is hardly a new phenomenon or invention. Magazines like the *Nation, Commentary, Commonweal,* the *New Republic, Harper's* and the *Atlantic Monthly*—to name a few bellwether journals—are driven by editorial content rather than advertising revenues. This is a critical distinction in publications—one insufficiently appreciated. Editorially driven periodicals have a long and honorable tradition in American society, while advertising dependent periodicals tend to survive only as long as the advertising base holds out. According to industry reports, there are roughly 800 magazine start-ups annually (using 1997 data), of which about 50 percent do not survive beyond the first issue. Nearly all of these are advertising dependent periodicals. The electronic periodicals may have fewer dependencies on advertisement and marketing skills, but they also have a much narrower band of viewers; sometimes reduced to friends or enemies of the bloggers.

Political periodicals which survive over a long time frame, offer testimony to the existence of an informed public interested in politics as a vocation. Whether such publications lead to better public policy is another matter. When any of these periodicals do falter, as in the case of *Harper's,* their pending demise or reduction of influence becomes a matter of public discourse, often followed by a bail-out of one sort or another. In the case of *Harper's,* the MacArthur Foundation did the bailing out. Such periodicals constitute extraordinary survivors from a host of publications whose names, much less orientations, are now only dimly recollected. Over the past few decades, a new, equally vigorous group of publications have emerged that rival older established periodicals. They share large-scale policy concerns but tend to focus on specific agendas more often than older periodicals do. These newcomers also tend to be more self-consciously ideological rather than rooted in public interest in general. the *American Spectator,* the *Weekly Standard, Academic Questions, Policy Review,* the *Independent Review,* and *First Things* ambitiously aim to shape America's consciousness on everything from education to religion by focusing more sharply on elites in academic life and the policy arena. The newer publications have an implicit belief that academics and policy makers constantly cross over one into another, and with social mobility largely determined by educational achievement and status, such a blurring is probably a correct assessment of the world of political publications as such.

Both newer and older periodicals tend to reach wider audiences through the commercial media. For example, major national newspapers, such as the *New York Times, USA Today,* and the *Wall Street Journal,* and major broadcast networks such as ABC, CBS, CNN, FOX, and NBC, rely on these periodicals as sources of information; the ideas put forth in these publications are thereby more widely disseminated. Indeed, the writers and editors for such journals are frequent guests as opinion makers on televised networks. They are selected more for the sharpness of their views rather than the intelligence of their opinions. But they have in common an independence from government controls and direction. That we have reached a point where such publications can no longer exist as rift-making centers in their own right, but must call for federal intervention itself becomes an issue in the politics of publishing—one that merits concern apart from the examination of the ideological and organization status of publications in their own right.

Authors and editors from political periodicals contribute disproportionately to "op-ed" pieces in large-circulation newspapers, and their authors and editors appear on television and radio talk shows, or as with the *Huffington Post* reports, develop their own model of electronic press reporting. They also contribute to editorial commentary in large-circulation, advertising-driven magazines such as the *New Yorker, Forbes,* and *Business Week.* Their editorial content is strategically placed in the *Congressional Record* and other such governmental organs of record. In short, publications of social and political opinion have a well-crafted, cultivated impact that reaches considerably beyond what their modest subscription numbers or advertising revenues might suggest. The trade-off seems to be that policy-oriented magazines, periodicals and online services gain much needed exposure, while larger-circulation media gain much needed legitimacy.

A number of larger, well-established periodicals have added a distinctive social science and social policy dimension. In publications such as *Time, Newsweek,* and *U.S. News and World Report,* findings of social science researchers are routinely reported and discussed—often with greater clarity and graphic precision than in the original reports. In addition, government agencies such as the Government Accounting Office or the Bureau of Management and Budget have expanded the word "accounting" and "budgeting" to include reporting on long-term social trends. Indeed, many GAO reports are themselves rooted in social scientific theory no less than social scientific data. These too are widely reported and appear in revised form in the editorial content of other publications. Together with political periodicals, social research organs have reshaped the delivery of political information to the American people. Through organs such as the Gannett-sponsored *USA Today,* the American public has come to expect that political views will often, if not always, be supported by rich information provided in precise graphical formats and summaries. The extent to which such data is received and reported second hand, without the benefit of political reporting as an act unto itself becomes a problem.

The purpose in organizing "The Politics of Publishing" was to bring together key editorial elites who have established a significant political base drawn from the American public. These individuals were asked to discuss fundamental issues that unite and divide them—although for the most part, the conference participants chose to emphasize the philosophical underpinnings of their own publications, rather than what distinguishes their periodicals from others in the field of political opinion making. Perhaps it was a sense of *noblesse oblige*, or simply a disinterest in what others are doing, but there seems to be a noticeable lack of concern for what others contribute, much less a sense of the common good.

Like science itself, democracy is rarely served by a universe of perfect intellectual agreement on contentious matters. Rather, such a gathering served as an effort to have people in the frontlines of magazine, book, and periodical life—whatever the format for delivery—exchange ideas and clarify the bases on which their specific publications are premised. Additional conference goals included examining impediments to growth—and even the risks of surviving in a mass media environment dominated by electronic rather than print services; and exploring the ideological assumptions and ethical foundations of explicitly political and social information networks.

In focusing tightly on these three themes the conference illumined—from the trenches—how influence is distributed, how information is disseminated, and how public opinion is formed in contemporary American society. Behind the rhetoric of a free press and free expression of ideas, there must be specific periodicals and publishers that exercise such freedoms. And despite constant arguments to the contrary, based on theories of benevolence of the powerful (it is hard for me to believe that monopoly controls, in which intellectual market share is tightly concentrated) they are an ideal environment for the achievement of the lofty aims and claims of democracy.

Bringing together this special group of opinion-makers and idea generators into a single conference center in and of itself made for an interesting dialogue, one of worth and possibly of interest to a broader community of scholars and public opinion-formers.

This is by no means an isolated endeavor. Everett Dennis and his colleagues at The Freedom Forum and now at Fordham University graduate center have pioneered in developing self-awareness, as well as public awareness, of the role of media in our lives.

The key "ethnographic" findings that emerged from the conference were informal, but nonetheless, consequential. A summary of discourses presented a formidable challenge in its own right. Among the critical observations that became manifest during the conference were the following.

At times key players in these political periodicals have personal, no less than intellectual, differences. These can be deep enough to prevent people from cooperating, even when they share common beliefs and identical ideologies. Most definitely, there is no conspiracy of periodicals aimed to forge a political consensus. In this sense, political similarities are more often thwarted by psychological and personal differences than simple identities in the public square.

Editors and contributors to political opinion journals are not in the habit of making explicit the assumptions and presumptions with which they operate. Ambiguity can be a positive instrument for dealing with diverse and rambunctious writers and authors. The presentations tended to soften hard-line positions. Editors did not see themselves in ideological terms so much as occupying a niche within a broader political framework. In this, one senses that in political as in other forms of serious writing and editing, professional standards are extremely important, and often override ideological positions.

Editors of politically sensitive periodicals have different objectives. That is to say, some want a highly structured environment and a corresponding set of marching orders from ownership and managers. Others prefer freedom of action, exemplified by looser, non-binding formats. All such opinion makers are keenly aware of the policy implications of the editorial positions they take. They seek to do so in contexts of high standards of rendering information in a superior format to purely political propaganda.

Different periodicals represent different ages as well as distinctive traditions. Journals that are fifty to one hundred years in existence tend to stay with conservative, classical formats. Those in their formative years tend to innovate with design elements—whatever their political or policy perspectives. And those who service electronic audiences are far more likely to utilize exposé methods in the delivery of stories and ideas. In large part, audience differentiation determines strategies of delivery of information. Just how long older publications with a substantially ageing constituency can withstand the thrust of newer technological services, or what the blend will be is yet to be determined.

Some editors are close to their proprietors and are aware of the problems of ownership. Others are basically remote and even in adversarial relation with management personnel. As a result, different people speak at different comfort zones, reflecting a greater or lesser sense of organizational authority as opposed to personal authority. These differences cannot be characterized by political orientation. One finds dismissals of editorial personnel at the *New Republic* in the liberal camp and the *National Review* in the conservative camp. Much depends on the ideological commitment and force of the ownership. In the case of such manifest agenda-driven publications, whatever the format, this sense of overseeing the ideological parameters of one's publication or blog remains a very high priority. The readers select an ideology rather than choose a publication. Editors are very much alert to such shifts in orientation, since it impacts the size of an audience, no less than the contents of the message.

While political journals with an ideological mission are a far cry from those concerned with scientific exactitude, they are nonetheless careful not to permit ideological considerations to weigh so heavily as to admit or even foster outright falsehood. These political periodicals take seriously a tradition of honest reporting, albeit within the dynamics of partisan viewpoints. The problem is always the relationship between the truth of a position and the value of an ideology. In this, the nature of truth itself is an omnipresent concern for political no less than scientific periodicals.

The most serious problem to emerge in recent years concerns the placement of the political journal on the Internet. For the most part, editorial personnel are anxious to see their specific message broadcast as widely as possible. Concerns for the profitability, or even survival capacity of these limited subscription publications appear to be quite decisive. As a result, technological considerations are viewed differently: by editors and writers as a great opportunity to spread the word; by proprietors and investors as a serious threat and risk. But with new modes of accommodation, the interests of editors and owners seem to be slowly reconciling; although the final outcome of this plethora of products on the Internet and in cyberspace is still in doubt and in formation.

While accessing information electronically is simple enough, with a modest amount of capital required, in order to convert simple appearance online to generating an audience base requires powerful technological platforms and services. These can be costly, and involve personnel far less interested in politics than in job-creation. The engineering of ideas has outpaced the change in political values as such. And all who venture into the labyrinthine world of political publications seem aware of this disjunction of technology as a method from politics as the message.

The United States represents a peculiar national culture—one in which the conduct of politics is a relatively restricted activity to a *social problems* orientation. This is a society with strong proclivities toward psychological and economical explanations for key events. That is to say, its people feel deeply about social issues of personal morality and conduct, and then define their sense of personal worth by how satisfied they are with their work and by how much financial credit they can muster no less than what they earn. Affluence may be a consequence of growth, but that growth is uneven in terms of social sectors.

Such an environment owes a great debt to a long tradition extending from Thorstein Veblen to Daniel Bell. Our century can well be described as a series of unresolved contradictions between an impulse to produce and an instinct to acquire. But the priority we uneasily give the latter does little to lessen society's need to provide for its citizenry and address world affairs in a constructive way. As a result, politics in general and policymaking in particular have become intensely elite occupations. They are always conducted in the name of all the people, but they are also conducted by relatively small groups of people who are not always receptive to soliciting views from those in whose name they speak.

In a Platonic world in which decision making is critical to all of us, yet is based on the opinion of the few, the desire, and the demand of organs of public opinion to be counted in this process is very great. Thus it is those smaller publications that wield enormous, even inordinate power. Their influence extends far beyond the number of their subscribers.

Many publications suffer unduly from a labeling process. As a result, specific periodicals are read more from their adhesion to some anticipated "line" than for the specific content of an article or an issue. Indeed, many editors report that when articles deviate too sharply from expected norms, readers become angered and disturbed. This problem extends to serious publications in the international relations field, perhaps along with social welfare concerns, the most volatile in the area of policy-relevant materials. Thus *Orbis, Foreign Affairs,* and *Foreign Policy* are considered "right," "center," and "left" publications respectively, although in fact one can throw a blanket over all three and, with some noteworthy exceptions, the articles that appear in one could easily fit the mode of the other two. This overlapping content seems to have little impact on public perceptions of a journal, since such appraisal is formed on the basis of organizational and institutional sponsorship rather than actual editorial content. As a result, whether such high-level journals of opinion actually sway political voting seems quite speculative. Differences that exist tend to be small, whereas similar standards of evidence seem prevalent in each of them, and these standards are far higher than those of journals or magazines of opinion altogether.

This type of overview might well be challenged on a variety of grounds. It is, after all, an impressionist reading of public as well as political opinion. Further, it does not attempt to assess various commercial and financial elements that go into publications. But if this analysis is the condition of the political culture in which we find ourselves, it behooves us to take a serious and continuous look at the editorial role in the politics of publishing.

Political publications whatever the delivery mechanism, are more concerned with mobilizing their own advocates than with seeking new converts, or reaching a broad audience. Editorial personnel are often preaching to the converted, or at least, were offering live ammunition to the converted, rather than serving broader scientific purposes of giving "all sides" to different social and political issues. As a result, marketing is directed at those who share the premises and policies of the magazine and not to a general public, such as what *Time* or *Newsweek* might seek. This is also a corollary of the distinction between general publications aimed at an advertising base (and therefore requiring numbers to justify their charges), and specialist publications with a strong editorial mission, with little regard to marketing and advertising. For the political publication is able to survive precisely to the extent that it satisfies some constituency needs—that may or may not have a broader base. What remains is a mutual sense of frustration: the general magazine carries little political weight, and the ideological magazine carries great weight, but lacks a reader base or outreach. In this, specialist political organs of opinion are involved in a search for relevance often thwarted by the very marketplace they seek to serve.

Editorial personnel in these mission publications work for far less money than do their commercial magazine or journal counterparts. They are part and parcel of the belief system of the publications, and hence internalize the experience and political mission of the publications for which they labor. But this sense of sacrifice gives them a feeling of being close to the decision-making process. This is a far different approach than commercial editors have to their management groups. The editors of policy-driven periodicals are often true believers. Even when they are cynical, they perform within a range of views to which they subscribe. This absence of a wall of separation between editing and managing, indeed the reverse, a strong consensus between the two functions, especially in electronic formats, can cause volatile outcomes. Editors can threaten to leave if management accepts advertisements they find unpleasant. Managers can threaten to do the editing themselves if editors step too far out of line. Given the volatility, one must be amazed at the long tenure of most editorial personnel of political publications.

But for the grace of a relatively small band of publications, editors, and readers goes the practice of the free press. It is all well and good to constantly repeat as a mantra the "right" to a free press, but it is quite something else to *practice* such freedoms on a regular basis. The views expressed in these organs are occasionally outrageous and at times just plain incorrect. But if such intellectual mood swings or fashions were disallowed, or even discouraged, the thin thread that ties the world of politics to goals of ethics would be severed—to the durable loss of a democratic culture. The most pressing issue facing these journals is the transformation of a technology based on the print media to one based on Internet and electronic communication. For this promises to reduce an already small cluster of serious political publications even further; for the extent to which televised opinion can displace the print opinion is one that will be settled in the long run. The broad issues will shift from the political ideology to the political institution. For now, the issue is less the survival of such publications, but rather whether private, philanthropic or public agencies will keep them afloat—and with what ends in view.

Note

These remarks, updated to reflect the expanding power of non-print media services, are based on a conference held on "The Politics of Publishing" on December 17-18, 1996. In attendance, though not necessarily reading prepared remarks were individuals from the magazine-journal world; professional and scholarly publishers; policy and research institutes; and mainly social scientists from colleges and universities. In order to protect the anonymity of these people, only their affiliations are herein listed. And while my main emphasis was focused on magazine and periodical representatives, many of the views and opinions of people from other fields were taken into account to the extent that an informal environment makes possible. It should be added also that the conference was not only about a select portion of the media, but elicited support from a related medium-television. The event was covered and fully broadcast (and rebroadcast) in January 1997 on C-SPAN and C-SPAN2. The director of programming informed me that this was the first time that a conference held at Rutgers University had been so widely covered on campus.

References

Lewis A. Coser, Charles Kadushin, and Walter W. Powell, *Books: The Culture and Commerce of Publishing* (New York: Basic Books, 1982), 411 pp.

Gordon Graham and Richard Abel, *The Book in the United States Today* (New Brunswick, NJ and London: Transaction, 1997), 269 pp.

Irving Louis Horowitz, *Communicating Ideas: The Politics of Scholarly Publishing.* (New York and London: Oxford University Press [revised edition, 1991]), 329 pp.

14

Monopolization of Publishing and Crisis in Education

While tendencies toward oligopoly have been longstanding in publishing, the emergence of new forms of concentration in both the quasi-private and university sectors, more then offset tendencies toward autonomy and growth. While it is a statistical fact that more firms and more titles exist in 2010 than ten or certainly twenty years ago, the publication of thousands of titles in scientific and humanistic disciplines disguises their growth of monopolization and concentration. Mergers and acquisitions have been accelerated by virtue of the technological transformations in all phases of the publishing industry. One can count on one hand, perhaps two of those of generous spirit, the number of scholarly journals that are not enveloped by a few firms in Europe and North America (often a combination of the two). While the book industry has been somewhat less swallowed by this process, it too has shown a tendency to have smaller firms become enveloped by large conglomerates, in which only the name, not the vibrancy, remain.

A new condition prevails that threatens any equilibrium of the absorption of the old and the emergence of the new in the publishing industry. We now confront not only the pandemic nature of the new forces of monopolization, but a total imbalance of market share. Concentration has affected not only firms but dollar amounts to the point that the top ten firms now account for 80 percent of publishing revenues. This means that the literally thousands of smaller firms account for less than one-fifth of the total revenues, and probably no more than 5 percent of the profits. The amount dedicated to scholarly publishing is fractional in monetary terms, but substantial in cultural terms.

Let me turn then to the specifics of what I have termed the crisis in scholarly publishing—one that is epidemic not only in the United States, but transnationally as well. We can indeed start with the fact that formerly independent British firms, such as Routledge & Kegan Paul, Tavistock Publications, Croom Helm, Carswell, and Eyre & Spottiswoode—among the great lists in social and legal research—have been absorbed by Methuen, which in turn is part of Associated Book Publishers Ltd. And now, most recently, ABP was purchased (in a "friendly" takeover) in late 1987 by International Thomson Organization Ltd., headquartered in Canada. But this monopolistic condition is at least still held in check by a parent corporation commitment to publishing and information. In the United States, The Free Press, Basic Books, and Praeger – among others exist in name only, as part of monopolization.

The situation in the United States is somewhat more ominous. What one finds is the absorption of major book publishers by mass media. Thus, in 1987 alone, Scott Foresman was acquired by Time, Inc., Doubleday by Bertelsmann AG of Germany, Harper & Row by News Corporation, and Technical Publishing by Cahners Publishing. But the most serious development has been the absorption of Allyn & Bacon by Prentice-Hall, which in turn was acquired by Simon & Schuster, which had been previously acquired by Gulf & Western. This series of transactions reveals a more ominous pattern, in which essentially nonpublishing concerns are absorbing publishing operations, redefining them into either components of "information networks" or "entertainment centers."

The process of monopolization has not only changed the rhetoric, but threatens the character of publishing—away from the production of knowledge and toward more amorphous categories such as information and entertainment. "Market-driven" considerations are used as *prima facie*, hard-boiled rationales for changes in the publishing lists. Scholarly publishing is itself viewed as a medieval luxury best seen in philanthropic terms and best serviced through university presses. The problem with this argument is that university presses can service only a relatively small portion of academic production. The underlying substance is that commercial publishers that are wholly owned corporate subsidiaries cannot engage in activities of marginal return on investment. Into this cul-de-sac falls the unwitting researcher or scholar.

The new monopolization has a direct bearing on the ability of publishing to satisfy a fundamental constitutional guarantee: free speech. It is not that the leaders or the publishing industry lack liberality in any pedestrian sense of that term, but rather that pressures toward profitability in the new mega-corporate environment inevitably subvert independent decision-making. For example, larger monopolies are quite willing to join in the boycott against an apartheid South Africa, but quite unwilling to struggle for the continued distribution or production of basic texts in the name of ensuring a continuous flow of anti-apartheid literature to South Africa.

A free speech environment is more subtly eroded by different notions of appropriate profit goals that occur in large and small firms. The need for a return on investment or profitability means different things in big corporations with obligations to large numbers of shareholders than it does in small independents. As a consequence, new titles with a market potential of under 5,000 (or even 10,000) copies, may be defined as unpublishable, not by virtue of their content, but simply because the market potential is estimated to be too small. The larger needs of the parent firm overwhelm the characteristics of smaller publishing subsidiaries. Again, the unsuspecting scholar is placed in a cul-de-sac, from which he or she can hardly emerge unscathed.

Specialized publishing once enjoyed potential grant support. But even these supplemental funds have eroded with the greater demands from a larger pool of scholars for a diminishing supply of money for publishing purposes. As a result, the monies that are available to support such publishing often reflect the agendas of special interest groups. Thus, the scholar is faced with going into the information or entertainment business in order to gain publication, or limiting dissemination to friends and associates who can be sent copies of a report or a monograph directly and without benefit of the value added by services of a serious publisher (such as refereed reports, editing, marketing, warehousing, distribution, and advertising).

In fairness, it must be noted that many of the publishers absorbed into this monopolistic environment are thus assured of continued existence. Many imprints still exist, but the wine turns sour and the mix is different under corporate auspices. In place of research monographs the demand is for yet more basic texts, handbooks, and reference works. In place of the work with strong views and firm intellectual identity, the demand is for books and essays that are tailored to satisfy impersonal or noncontroversial needs. In place of long works with careful notation systems and complex charts, the demand is for brief, easy-to-read books, in which the scholarly apparatus is severely curtailed.

What can the individual scholar do about the current situation? To start with, one must accept the reality, perhaps even the irrevocable nature, of this monopolization process. Thus calls for congressional review of industry-wide concentration are likely to fall on deaf ears. After all, in this environment of three automobile manufacturers and five major pharmaceutical houses, the likelihood of legislative relief in an industry that boasts of a "big ten" will not easily be heard. Similarly, it makes little sense to demand that university presses do much more. They are already overburdened with monographs and undersupported by university funds. Such calls for philanthropic relief represent an abandonment of marketplace approaches and techniques as well as a surrender of economic self-sufficiency in the area of academic publishing. What then can be done to alleviate this crisis?

First, every major grant for research purposes should have funds built in for publication and dissemination. It makes no sense—to have hundreds of thousands of dollars allocated for research by major granting agencies, while few if any provisions are made for support of publication. But, as anyone who has applied even for typing support soon finds out, granting agencies are particularly ruthless in this area. This would be easy to open access, but the library and scholarly community would be advantaged by such developments.

Second, institutes dedicated to specific research areas should start allotting a portion of their budgets to the dissemination of their findings. That means an increase of publishing in partnership. A whole range of possible forms exist for cooperative relationships between professional publishers and research institutes. At the moment, in a bureaucratic structure like that which exists in Washington, there is a virtual lock on materials produced. Few results are disseminated beyond narrow elites. Research reports gather dust even when printed, because little effort is made to make the market aware of their availability. Professional publishers know how to do just this.

Third, scholarly associations—especially newer, innovative agencies reflecting advanced trends in the sciences and humanities—must themselves become far more vigorous in the publishing environment. In the past, such associations have tended to leave book and monograph publishing to a handful of university and scholarly publishers, while themselves restricting activities to key journal publication. Professional associations must themselves enter more vigorously into a publishing mode. They must develop new forms of relations with specific clusters of scholarly publishing units to broaden the base of knowledge creation. Such an expansion into scholarly publication by scholarly associations would relieve the burdens of the researcher in a monopolized environment, and even modest success would provide a measure of safeguard against corporate takeover.

At stake is not the general problem of monopolization, not even the specific availabilities of scholarly outlets. Central to all of this, and what justifies the use of the word "crisis," is the threat posed by present tendencies toward mega-publishing to freedom of speech in a free society. If we are to avoid a dialogue of the deaf conducted by cautious elites, there is a need to secure a broader research base built upon close-knit participation by the scholarly community in the processes of information production and dissemination. The accelerated pattern of monopolization in the book and journal industry provides a window of opportunity no less than a specter of closure. Unless the universities and major research agencies of the nation embrace this opportunity, the problem of disseminating knowledge will quickly give way to the still broader problem of knowledge creation as such.

15

Publishing Responses to Economic Crisis

"Alongside the need to eat, be fruitful, and stay out of the rain, the need to know deserves an equal place. Publishing is chief among the ways we humans satisfy the need to know, and the nature of publishing has been changing since time immemorial. The difference now is the rate of change and the proliferation of options... If content written, edited, polished and marketed are to survive the onslaught of alternative forms of media, then it must create superior value for its producers and consumers alike."
—Knowledge for Generations: Wiley and the
Global Publishing Industry: 1807-2007 (p. 459).

We have it on the shaky authority of *The New Encyclopedia Britannica* that "the great trade slump that began in October 1929 brought a swift decline in the prosperity of American publishing." This was declared to be a worldwide phenomenon, since it was also the case that in Europe "sales declined, profits were negligible, and there were many bankruptcies." It is also the case that attempts to remedy a bad fiscal situation were weakened by tremendous antagonisms between publishers, booksellers, distributors, and authors. But this view is not quite the story of the past, and even less of the present. Those who are undergoing painful readjustments to a 2009 marketplace that is heir to boom years in publishing must now adjust to the second half meltdown in the 2008 marketplace. This is a complex condition open to a variety of conditions, and hence to a multiplicity of responses and adjustments.

That said, along with its *prima facie* damages to the society, the Depression brought with it enormous changes of a positive sort.

In sheer quantitative terms, the decade of 1929-1939 displayed remarkable resiliency. The number of titles remained quite stable, with declines in some years like 1933 but sharp inclines in 1935 and a leveling off at those higher production figures throughout. Profit margins were modest, but by an emphasis on everything from cost cutting to discount coupons to increase sales, many older firms survived, and some newer firms even thrived. There was a sense that publishing was a counter-cyclical industry, at least in part, and that the audience for good books was considerable. To be sure, the use of books as source material by the newly established "talkie" film industry drew heavily upon major authors and titles. This in turn stimulated increased production in all categories of books from romance novels to westerns and mysteries. And while the emergence of popular culture as a field did not yet transform academic publishing as such, the spillover effect was found in areas such as travelogues and ethnographies about far away places.

What is less well understood is how the publishing world coped with the macroeconomic condition in the Great Depression eighty years ago, and what lessons—if any—this cataclysmic event holds for the present epoch. For ours is a period in which publishers are not only confronted with an economic set of unforeseen circumstances, but also with a sea of technological changes very much unforeseen, if not uniformly, even less positively by all players concerned. I would like to briefly review that earlier Depression decade and try to examine if what came before could have been avoided, or at least, minimized with particular impact in terms of the pain inflicted on scholarly and professional publishing. Because of the remarkable effort by John Tebbel in his magisterial four-volume history of American publishing, this brief history review is possible.

1. *Paperback Publishing*. One of the great breakthrough elements of that period is the development of paperback publishing as a mass technique. While scholarly and university presses were slow to accept paperback publishing, fearing that it would cut into the review process at one end and revenues from expensive cloth editions at the other, it caught on nonetheless. From Blue Ribbon

Books in 1933 to Penguin Books in 1935 to Pocket Books under the aegis of The New American Library in 1939, commercial houses began to exploit the mass potential of inexpensive editions in paperback. What started as an inexpensive way to reproduce classics for the masses soon began to flourish as a way to put into production new editions of books previously published in hard binding. The Depression era force fed a consciousness that broke down class stratification and was a critical factor to growth and economic stability.

2. *Partisan Publishing*. While religious publishing institutions like Sheed & Ward were long involved in book titles that expressed strong clerical and ethical emphases, the emergence of radical groups on the right as well as left, were accompanied by the coming into existence of publishing houses dedicated to marginal political forces such as communists, socialists, and fascists. These groups gave expression to views that might otherwise been left in the lurch. Indeed, it was such presses that published the works of significant European thinkers from the physiologist Ivan Pavlov to the economic historian Oswald Spengler. Such publications from International Publishers to Stackpole Press were also engaged in bringing to America the looming sense of further global conflicts involving mass numbers and powerful ideologies. Such firms enlarged the scope of academic freedom and civil rights, albeit inadvertently.

3. *Book Clubs*. While The Book of the Month Club and The Literary Guild actually began in the late 1920s, they did not pick up momentum until the early 1930s, when mass marketing techniques became fashionable and publishers became supportive. They became important sources of building sales and reducing costs, along with smaller, specialist book clubs, such as those that reprinted classical works in splendid binding and catered to needs of specialized audiences. This sort of activity had the heavy backing of the major commercial houses, which began to see such targeting and twigging of the market as stimulants to their most popular works, and reached out to new audiences precisely at a moment when the marketability of even best sellers had slowed. The book club became a source of decision-making by publishing

elites aimed at the masses—itself an innovation. Senior editors, like Clifton Fadiman, became intellectual figures of note in their own right.

4. *Supplements and Review Sections.* The decade of the 1930s also saw the flowering of Sunday book supplements that featured cultural events as a unique part of the paper. From these emerged an emphasis on book reviews in many major newspapers. Some newspapers like the *New York Times, Washington Post, Los Angeles Times,* and *Chicago Tribune,* spun off substantial book review sections which became a source of advertising revenues from publishers and stimulated increased interest in books no less than increased interest in featured columnists from potential readers, authors, and editors. It is of no small concern that this pattern has been dramatically reversed in the first decade of the new century with many newspapers canceling its book review sections, or at least trimming back reviews with hardly a whisper or whimper. Indeed, the number of newspapers has declined in major cities, with a corresponding loss in revenues and arguably influence.

5. *Textbooks as a Major Market.* The 1930s also witnessed the emergence of textbook publishers as a mass industry unto itself. As colleges and universities began to expand considerably from elite to mass student bodies, from a 10 percent to a 50 percent demographic factor somewhat later, following World War II, targeted texts following course offerings came to be seen as a vital force. This was not necessarily for the sake of learning but for the need to graduate, for what has come to be known as upward mobility. The text market in turn spawned a pony market, summaries of classic work. The idea of classrooms as a coming together for learning whether in English or in physics gave way to demanding texts in areas that prepared young people for careers in every area—especially medicine, law, accounting, and the physical sciences. Class sizes became large as the demand for texts increased. These practical and scholarly materials were a ready substitute to hiring more teachers and instructors. The social sciences were not far behind, although they reached their own pinnacle in enrollment after World War II and the emergence of strict textbook publishing as an independent spin-off in the 1950s.

6. *Discovery of Europe.* American publishers discovered European writers during the thirties—in part as a function of the renewed threat of a world war, but also as a mechanism for expanding the fiction part of their lists beyond pot boiler romance novels, westerns, and mysteries. Alfred Knopf and Kurt Wolff in particular recognized a growing cosmopolitanism of the American reading public, and in 1930 and beyond responded to quite a new set of circumstances abroad and cultural transformations in the United States. Beyond the new acquaintance with great fiction writings from Europe, was a new familiarity with major works of social scientists abroad. Routledge & Kegan Paul, Allen Lane, and Weidenfeld & Nicholson among others—that is to say, European-based firms, began to treat the American market as a serious outlet for their works of primary sociologists, political scientists, anthropologists, psychologists, and economists lodged in Europe. While this was stimulated in considerable part as a function of intellectual and publishing migration to Great Britain from fascism, it had a profound impact on what constituted commercially viable projects in the 1930s. That global outreach in both directions never ceased from that point onward.

7. *Professional and Target Publishing.* Special developments took place in the world of scholarly organizational life as well as publishing. While the American Association of University Presses was chartered and informally assembled in the late 1920s, the actual rules of membership and standards of operation for university-based publishing activities took place a full decade later. It was the 1930s that gave a frame of reference to an entity known as scholarly publishing, as a unique activity unto itself, and distinct from commercial and text publications. Eight publishing units that were campus based mushroomed to seventy by the close of the thirties. This development also established a groundswell of publication units attached to professional associations connected with medicine, law, chemistry, and physics. Not long thereafter, fields such as economics, psychology, anthropology, and sociology established distinct arms of publication separate from, but allied with, both commercial and university-based publication.

8. *Niche Publishing Under the Radar*. New start-up publishers defied the Depression by multiplying far beyond what could have been expected. Not only were they enriched by the immigrants from European fascism, Nazism, and communism, but also by Native American firms that found niches undiscovered by the major players. From Zondervan and Protestant Publishing, Schocken and Jewish Publishing; New Directions, Crown, Duell Sloan and Pearce in serious (and dicey) fiction; and from Basic Books in 1933 to Praeger Publishers in 1938 there arose strong non-fiction houses with sharp political edges. It should be noted that while the multiplication of firms spawned in the Depression years, the rise of monopolization dissolved by the end of the century. So-called serious mid-list books were abandoned by major commercial houses, smaller publishers were absorbed by the monopolies, and most social science publishing was left in the hands of the university presses. Perhaps five to ten independent houses emerged in publishing on minority and racial themes and on gender issues. So the decade that started in a blaze of pluralisms ended in the still waters of monism.

This expansion driving a challenging period in American publishing history points the way to how the social sciences might thrive under adversity. It most certainly does not prove that harsh circumstances are a necessary or even a welcome prod to success in the scholarly publishing arena. But it does indicate that in addition to the standard reasons for success—such as sound business principles of management and financing, or rational systems of production, marketing and distribution, there are factors at work that should restore a sense of calm realism about the current publishing climate, which from the outside may appear alarming and just plain chaotic. Now let us turn to the present in comparing two periods in American economic history noteworthy for downturns rather than developments.

1. *From Inventory Size to Property Rights*. The quintessential fact of the present is a shift from the measurement of publishing in terms of the number of copies sold and on hand, to property, that is

to say the legal rights to publish and republish books of a serious nature. The emergence of electronic books and digitalization, or simply paying for and downloading books from electronic holdings for personal usage, has become a central fact of scholarly publishing. The speed and quality with which titles can be reproduced digitally has made the need to hold large inventory of books obsolete. On the other hand, it has created a strange decentralization in which printing decisions may be in the hands of vendors. How successfully publishers work in this new environment is a large element in defining how severe the depression will be. End users now expect to have books in their hands in hours or days, and not weeks or even months as in the past. The new information technology is a critical factor in defining the tensions laden in the present situation.

2. *From Electronic Processing to Final Copy.* While the final product publishers sell still remains largely concentrated on hard copy, the intermediate steps—from manuscript transmission to editorial and production stages—have dramatically shifted to electronic systems of everything from word processing to print publication. This is not a uniform process, nor is it one without pitfalls of its own; for in this shift there has also come a far more significant role of the author in the editorial production as well as the creation of ideas. As a result, everything from font selection, cover design, catalogue placement to royalty payments has become impacted by this new information technology. In its optimal form, new levels of cooperation between publisher and author replace old styles of animosities and antagonisms. The costs of production and the time it takes to deliver final copies may be reduced. These are extremely important in economically pinched times, but of no lesser significance, the author is more closely bonded to the publisher in the entire process of pre-publication elements and increasingly in the post publication process. In short, beyond a shift in formats, tools, techniques, standards, and implementation of a new technology, is the subtext of this transformation: a huge shift from inventory counts to proprietary controls. It is at this level that the pitfalls will be resolved.

3. *From Publisher to End User.* A critical byproduct of the electronic revolution is the weakening of intermediaries—from book-

stores to foreign distributors. I say weakening since the need for strategically placed individuals or firms that distribute and market continues to be a factor. That said, with the widespread infusion of HTML (hypertext markup language), XML (extensible markup language), and a variety of developments in progress related to connectivity of publishers and end users on their websites directly, profound differences have emerged in the publishing process as such. Indeed, the very relationships of authors and editors to publishers have shifted ground in ways unimagined even a decade earlier. Some authors have become entrepreneurs and command a good deal of attention in entirely distinctive ways. By the same token, with a reduced work load in scholarly publishing houses, the traditional costs of production have shown a marked downward turn. At the same time, publishers are incurring new costs related to preparation of work for dissemination in new electronic format. This needs to be understood to avoid struggle for sharing returns on investments. The idea of technology as an unmitigated good or evil is being replaced by a deeper sense of commonalities between publishers and end users (not to mention authors and editors), and also a disquieting realization that older models simply are archaic and counter-productive.

4. *From Mail Shots to Rapid Data Transmission.* Perhaps the most important aspect of electronic transmission is not so much in the final print product, but in the marketing and sales aspects of the book. Publishers now maintain and transmit electronic databases either to replace or augment hardcopy general catalogues in paper form. The costs of such activities, while superficially less than old fashioned models of catalogues and brochures, is often deceptive. Suppliers and vendors alike require technical information and up-grades that shift the burdens of management and warehousing from these service agencies to the publishers. There is the further risk that such external agencies call the shots, that is, determine what the ubiquitous "market" can absorb in terms of buying and selling. This is more of a problem with larger commercial publishers than with the world of scholarly books. But it is increasingly becoming an issue for these smaller houses as well, since demands for profitability increase as economic conditions deteriorate. That said

the rationalization of distribution is an immense shift that provides for unparalleled opportunities for publishers of specialist works to compete on favorable terms with their larger compatriots.

5. *From National to Global Publishing.* The ability to outsource all manner of tasks in the publication process has allowed for a new division of labor. Compositions, design, printing, and binding have often migrated to nations in Asia that have provided competitive price structures, and maintained relatively high standards of the final product. Post-publication conversion to electronic products is also being increasingly performed overseas. As a result, prices have remained relatively stable, and costs kept in check amidst spiraling upward costs of supplies. But here too, the publishing industry has incurred serious losses. Outsourcing involves loss of controls with respect to the final product that no amount of instructions and system verification can overcome. Standardization of the book in all its aspects replaces variations that make for uniqueness.

The boutique elements in specialist publishing particularly are replaced by commercial elements of "gang runs" in printing, and reduction in composition and design elements in editing. Even the size of the book is increasingly determined by what can be done efficiently with current technologies, and less by what a book may actually require for maximum intellectual returns.

6. *From Authoritarian to Democratic Modalities.* One of the most dramatic changes is the creation of a level playing field of searching for titles, articles, and chapters, as a result of such search engines as those created by Google, Yahoo, Microsoft, and others which also provide comparative information about price and cost of such works. The quality of the product, the precision with which any given title fulfills a need, and the rapid availability of the book in the hands of the end user replace older models in which the monopoly publishers controlled everything from the printing schedules to the delivery dates. This is particularly important to the world of university and scholarly publishing where college- and university-driven subject areas determine need, rather than publisher preferences.

If there is a subtle factor that determines survival and growth in the present era it is this shift from monopoly definitions of needs to professional and policy needs in the problem solving arena. Once again, the individual scholar has a much increased role in outcomes. The strange anomaly is that in an era of impersonal, electronic publications, the individual, far from fading away as a factor in decision, or being reduced to a puppet of marketing whims, becomes critical. With this comes the need for a level of publishing intimacies with the end user thought to have vanished at the close of the nineteenth century. These new relationships may not be one of author to editor, so much as professional to publisher in their various guises and responsibilities.

7. *From Long Term to Short Term.* A critical difference between our moment in time and earlier periods in the history of publishing is the reduction of time intervals: everything from decisions about manuscript acceptance to the publication process itself has been impacted by the computer age. And while academic and scholarly publishers are arguably the slowest to appreciate this change, it has brought all concerned deep into the new century. In a period of severe economic meltdown, such issues as turn around time in editorial decision making to manuscript production become crucial. Time is perhaps the most important single variable for an author, more so than royalty rates and house prestige. Thus even referee reporting, perhaps the most recalcitrant element in the shrinkage of time for publishing decisions, are showing signs of awakening. The post publication ability to disseminate materials online and in electronic formats are very much part of this revolution in publishing time. The ability to survive an epoch of economic downturn may well rest on the management of time.

8. *From Lone Wolves to Cooperative Association.* Publishing—especially academic publishing—has in the past been a relatively inexpensive operation in terms of basic costs. It was once said that the entrance price of a scholarly publishing house was just about equal to that of a new automobile. But such an era was predicated on a variety of factors that no longer exist: above all, the absence of competitive firms in the small presses. At this point in time, the

technologies needed to drive the modern publisher and to get the publishers' works into distribution channels have raised to sub-stantial proportions. They are fueled by large publishing houses at one end and individual scholars who want boutique treatment at the other. In this environment one can anticipate the emergence of cooperatives as well as amalgamations of smaller units into larger combines. The need to share costs and take advantage of new developments in a rapidly changing technology, are driving smaller publishers into such arrangements. And in a depression-like atmosphere, one accompanied by technological opportunity, and hence quite unlike what took place in the 1930s, the answer will be further cooperation and amalgamation—especially in terms of distribution as well as creation of information. Survival once again will depend on the ability of scholarly houses to cooperate at the technical level while continuing to compete at the intellectual level—no small feat, but a necessary proximate goal.

We should firmly keep in mind the specific advantages no less than general maladies that confront us. Academic publishing is to a large degree counter-cyclical. That is to say, our lists thrive in conditions of adversity precisely because people, especially professionals who are responsible for administrative chores and decision-making policies, seek answers to quotidian concerns in books and journals, rather than relief from the cares of the world or mysteries that deepen the problem rather than aid in problem solving. Thus it is those close examinations of recent data that indicate contrarian movements: some areas and fields, especially those that are purely involved in competition for the entertain-ment dollar may suffer setbacks, while those engaged in moving the society from square one to resolution two may find a readier acceptance for their wares than might otherwise be the case. In this sense, the present condition is an opportunity to expand the democratic horizons of our area of publications. It was the case in the 1930s, and it should be the case in the coming years as well.

However, it is worth noting that in the larger scheme of things a certain huge difference is manifest. The first depression was one that was driven by sociological factors. These external, larger elements—from the rise of mass higher education, professional societies and policy agencies, to governmental expansion into the affairs of what formerly was considered private matters, all served to enhance the place of book publishing. Intelligence was, of course, required, but opportunities were readily available. The situation at present is far more internally driven, which is a function of technological factors that have placed the printed word front and center, but also at risk. While there are overlapping forces at work, this qualitative distinction can hardly be minimized. The current situation is one that offers opportunities as did the past, but it also opens up risks to the likes of which were unheard of ten years ago, much less eighty years earlier. How these new technologies are managed, and above all, integrated into the tasks of the expansion of knowledge and its integration with values will determine the results.

In this connection, one must reiterate the importance of sound business practice. This means maintaining a careful check on new expenditures—as needed not as wanted, balancing accounts of receipts and payables, so that basic obligations are met, from salaries to staff, taxes to state and federal governments, paying invoices to vendors, and royalties to authors and distributors. These are simple things to say, but somewhat more difficult to execute in an everyday working environment. Having come to maturity in an academic environment, each of the above represents not simply obligations to others but a learning experience to myself. In good times or bad, these are obligations that must be met. The danger is exaggeration, a belief that "leveraging" and borrowing can somehow jump start a publishing house—as if throwing money at a problem is a panacea that always works. The opposite danger is fear, a belief that cutting back on staff, dismissing personnel because newspaper headlines put forth such events, is a solution. Maintaining a balance, walking the walk, is a tightrope activity. But it is that sense of balance between risk and prudence that will see the good publisher through the hard times. And the good publisher is one that publishes good books!

Having stated the obvious, it is also important to note that the readers are more critical and selective than in the past. People are less likely to accept ideological positions that are lacking in grounded evidence than they were in good times. They are less persuaded by the quality of rhetoric and more by the touch of reality. They do not buy books to confirm what they see on television screens or blogs, but to move beyond such generalizations into the particularities that impact specific industries and the human conditions. The challenge of remaining true to the needs of professional and scholarly publishing in troubling economic conditions is not to be minimized. But they are the challenges that must be confronted, not surrendered to. These remarks are admittedly tentative and impressionistic. They will be confirmed or disconfirmed by hard data yet to come, and by new technological developments that remain to be identified. But somehow, I believe that the industry in which we share our common struggle will thrive. The issue will be centered on who the survivors are, not the continued growth of serious publishing as such. Of that much I am confident.

References

E.M. Crane, *A Century of Book Publishing, 1848-1948* (New York: Van Nostrand, 1948), pp. 68-179.

The New Encyclopedia Britannica (15th edition), "Publishing" (Chicago: University of Chicago, 1993), Volume 26, pp. 415-449, esp. pp. 427-428.

John Tebbel, *A History of Book Publishing in the United States. Volume III: The Golden Age between Two Wars, 1920-1940* (New York and London: R.R. Bowker Company 1978), especially pp. 427-618.

Peter Wiley, Susan Spilka, and Barbara Heaney (editors), *Knowledge for Generations: Wiley and the Global Publishing Industry: 1807-2007* (New York: John Wiley & Sons, 2007), p. 459

16

Publishing as a Vocation:
The Necessity of Independence

*"There is not one individual who is not the pos-
sessor of dear and cherished unpopular convictions which common wisdom forbids him to utter.
Sometimes we suppress an opinion for reasons that
are a credit to us, not a discredit, but oftenest we
suppress an unpopular opinion because we cannot
afford the bitter cost of putting it forth. None of us
likes to be hated; none of us likes to be shunned."*
—Mark Twain, The Privilege of the Grave (1905)

I choose to open my remarks with Mark Twain for reasons that
I hope will become transparent by the close. But the deepest debt
I have, and this will become apparent, is to the brilliant lectures
by Max Weber on "Politics as a Vocation" and "Science as a
Vocation." Many have read these remarks; few have been able
to implement them fully—not necessarily for want of trying so
much as for lack of opportunity. It has been my great fortune to
link the two. What recalls these two essays, beyond their pellucid
style and quality, are the concerns that make the social sciences
a *Beruf*—something more potent than a vocation—a calling with
an inescapable moral edge, applied to a world more of obligation
than of joy.

The expression, not the suppression, of unpopular opinion is
precisely what publishing at its best is about; that goes for all publishing. However, scholarly publishing poses hurdles and problems
of a unique sort; these are well known but rarely spoken of.

What follows will not be labor vague generalities about free expression of ideas, or celebrate the existence of diversity in the academy, or even the virtues of independent publishing houses. Rather, I shall take up moments or case studies in the history of Transaction that measure such generalities, and put to the test just how diverse our society is in the face of strongly felt challenges with sharp moral edges. It was Johann Wolfgang von Goethe, in *Faust*, who first reminded us that the struggle between good and evil is relatively simple and straightforward; it is the struggle of good versus good that is so painful and difficult to dissect. Publishing is a powerful test of Goethe's deep insight. The recognition of the costs and benefits of a free publishing world is a time to reflect, not an occasion to celebrate.

I

The first such moral "test" we faced at Transaction after initiating our book and journal publishing program took place in 1973. Our first publication lists were clearly dependent on people and subject areas that were developed in the first decade of publications of our social science magazine, first known as *Transaction* and later retitled *Society* for a host of solid reasons. These are not however, relevant to our present concerns. The earliest titles emanated from special issues in *Society* and dealt with the wars then raging: in Vietnam, in the ghettoes, and in the academy. The Middle East was also on the agenda, not least the latest (1973) round of struggles between Israel and its Arab neighbors. That final element was evidently what alerted our very small firm to the very large and powerful Wiley Publishers.

I received a communication from the legal counsel at Wiley that they were interested in the sale and transfer of inventory of thirty-odd titles it had published on behalf of the Shiloah Institute for Middle Eastern Studies in Israel. It was made clear that a modest sum would be involved if the transfer and sale could be made quickly and quietly. To be sure, my recollection is that we met at the home of the Wiley officials in northern New Jersey, rather than at their editorial offices in mid-town Manhattan. The deal was struck quickly, with an amazing lack of administrative or bureaucratic concerns. Needless to say, the reasons for this fire sale were puzzling—especially since the titles involved were written by world-class scholars in history, law, and social research. Indeed, both in design and intellectual content, they were books considerably in advance of our technical capacities at the time. This only added to the mystery of that moment.

At the meeting with the Wiley attorney, the key representatives of the Shiloah Center at Tel Aviv University were present. They made clear the reason for the sale, although I suspect that they did not look closely at the painfully limited capacities of Transaction at that time. It turns out that the Soviet Union had long-standing agreements with Wiley for the publication in English of basic works in physics and mathematics by its leading figures. Wiley, as publisher of Inter-Science, was a primary publisher in these fields, along with McGraw-Hill in the United States and Great Britain and Springer Verlag in much of the rest of the world. Copyright authorities in the USSR were instructed, presumably by officials within the Communist Party, to inform Wiley that they were under severe pressure from its Middle East associates to observe an informal, unwritten, but hard-as-nails boycott of Israeli products—including literary and scientific products.

In a nutshell, Wiley had to choose between lucrative activities in science and technology with its Soviet counterparts and the relatively modest list of titles published on behalf of the Israeli institute. Knowing the Wiley management well, I have no doubt that the decision to divest the Israeli titles already published, no less than titles they had signed up for yet to appear, was a painful event. But it was also clear what the choice would be, and the axe had to fall—Soviet scientific writings, not Israeli anthropological monographs. I suspect that Transaction was not the first publisher contacted to take on the Israeli titles, but we proved to be the most rapid in decision making and asked the fewest questions. It must be noted that this was a tremendous boost to our publishing program: we were able to go to the Jerusalem Book Fair in 1975 with a powerful representation of outstanding books in and about the Middle East, without having to absorb publication costs at least ten times the sum we paid.

The results were transparent: Wiley relieved itself of a nettle-some burden, Transaction acquired a list of titles it could promote with ease, and the Soviet authorities could provide satisfaction to their Arab Middle East associates that they were acquiescing in the boycott of Israeli titles—just how enthusiastically I never could detect, nor did I much care. But if the fiscal arrangement worked to the general good, it also served to highlight for me the general evil: the admixture of politics and publishing that one entered the academic fray precisely to overcome! All parties to this minuscule deal (minuscule to Wiley, at least) were satisfied; none, apart from, Transaction, I suspect, were particularly happy. Not even the Soviet government, which did not take kindly to acquiescing to outsiders, found this palatable, even if they shared a common ideological posture at that moment. The players were more victims of global politics than villains, or, for that matter, heroes. At the same time, I admit to a certain pride in the fact that Transaction was well represented and fêted at the Jerusalem Book Fair of 1975, while Wiley took a pass.

II

The next large-scale challenge and response, to borrow Sir Arnold Toynbee's famous phrase of the dynamics of history, came in the mid-1980s. Transaction at the time was represented in the English-speaking world outside North America by Holt-Saunders—itself an amalgam of American text and British medical publishers owned by CBS. This was a good situation for a small firm like Transaction, since Holt-Saunders reached parts of Asia such as Singapore and, more importantly, Anglophone Africa. It was in this periphery that the problem was manufactured.

We received word from the State Commission of Education in Michigan, and I believe several other state boards of education, that any publishing house with representation and trade to South Africa would be denied the privilege of sales to schools in that state. We were informed that a boycott was in effect and that Transaction, by virtue of its Holt-Saunders affiliation, was on a list of those violating the boycott. Given that this was a time when commercial houses had abandoned African and African-American studies as unprofitable, whereas our house had become a beacon for black American scholarship—thanks in part to a conscious decision and in part to our social science policy orientation—this represented a potentially serious interruption and loss of revenue, as well as a loss of sanity.

To be sure, the Michigan boycott edict had as its main purpose the major textbook houses. Sales to high schools and colleges that followed the lead of state and local boards of education represented large money for the "majors." Within a very short time, announcements were posted that McGraw-Hill, without so much as a whisper, would pull out of South Africa and, moreover, sell its interests in the local markets to a Johannesburg publishing house. But a funny thing happened on the way to this particular forum—not every publisher was willing to stop selling to South Africa. I wrote two pieces, one for *Publishers Weekly* and the other for the *British Bookseller*, rejecting the boycott approach to apartheid as woefully and shamelessly counterproductive.

It turned out that Nelson Mandela himself stepped into the controversy, one in which the American Association of Publishers and a few others exercised "quiet" diplomacy, and prevailed on the leader of the African National Congress to issue a public declaration exempting serious publishers from the boycott and making it clear that good books from the West were very much in demand and much needed in the struggle against racism and the apartheid regime. This was welcome news, to say the least, and the various state boards that had signed up for the boycott quickly evaporated. But the damage was done. Major text houses had a difficult time re-establishing broken links to South Africa, and other firms, including our Holt-Saunders repositioned themselves in the marketplace—although I hasten to add that there is no direct cause and effect between the boycott and such organizational changes.

If the interference of the USSR in the normal supply of, and demand for certain books served to highlight the role of global politics in the conduct of the life of the mind, then the role of local American educational bodies serving as agencies to determine book shipments and sales was the lesson of the attempted boycott of books to South Africa. That it failed was less a function of the publishing community—indeed, with rare exceptions, that community remained mute about the risks of such actions to the world of learning as such—than of Mandela's courage in bringing closure to such nonsensical feel-good politics. While I was proud to speak out in public on this bizarre incident, the timidity of even monopoly giants to make a public statement or offer personal support was an unnerving reminder of just how tenuous the notion of a free press is, even in a free society.

III

The third case study of the fragile nature of scholarly publishing concerns the history of the Grand Mufti (al-Husseini and others), which provided material as well as ideological support to the Nazi regime during World War II and was particularly virulent in its belief in the value of the "Final Solution," that is, the death of the Jewish communities of Europe that resulted in the Holocaust. *The Icon of Evil: Hitler's Mufti and the Rise of Radical Islam*, authored by Stanford University professor David G. Dalin with John F. Rothmann of the University of San Francisco, details the tradition of such figures not only in the past but as prototypes embedded in the leadership of Hamas and Hezbollah in the present epoch. It should be noted that this book was published by the distinguished commercial publisher Random House. It was a modestly successful launch—at least by academic standards—but in late 2008, several months after publication, the first printing was sold and the book quietly removed from further marketing or promotional consideration.

Random House licensed the work to Transaction for the munificent fee of $100.00—lock, stock, and barrel. We were, and remain, delighted with this acquisition. Gift horses from mighty commercial publishers do not come along often. The new Transaction edition has an introduction from the Harvard professor and attorney Alan Dershowitz, and, however the book does in the marketplace, I can assure would-be suitors that no fire sale in the night will again take place. It is clear that a thread runs through a number of these case studies: from the ceaseless interest in the Jewish Question, further deepened by the emergence of the Israeli

Question (known in some quarters as the issue of Zionism) at one end, to issues of social welfare and social deviance at the other. We did not seek out such special emphasis, we did not attempt to sensationalize issues of public concern, and we did not capitalize on media potentials or pursue political gain of any sort. But the struggle for a free and independent publishing world creates its own momentum, and serves to separate the stragglers from the strugglers.

Transaction is not a religious publisher, or, for that matter, a "Jewish House." It certainly is not one that specializes in books on deviant behavior. Yet it is clear that such a linkage does exist in the public perception and cannot entirely be avoided. It is part of the horrid tradition of anti-Semitism that links between Jewish beliefs and social perversities that are made routinely and as a matter of fact. The publication of such books as *The Icon of Evil* or, as we shall see below, *Tearoom Trade* will hardly lead to an abatement of such canards. That said, it is the great tradition of English and American publishing after the rise of fascism, Nazism, and, yes, Communism that Jewish-owned or -managed firms became innovative bellwethers of social science and cultural studies in the West. While this is not necessarily a matter to be shouted about, neither is it one that can be shunted aside. Indeed, Transaction has recently published a solid collection—*Immigrant Publishers: The Impact of Expatriate Publishers in Britain and America in the 20th Century*, edited by Gordon Graham and Richard Abel—that includes such extraordinary pioneers of talent and courage as George Weidenfeld, André Deutsch, Kurt Wolff, Frederick Praeger, and Ursula Springer, among many others, who understood and acted upon a book and journal publishing environment that lagged far behind its media compatriots in the film, newspaper, and magazine industries. It is within that unique, albeit unfortunate tradition that our edition of *The Icon of Evil* becomes the mark of publishing courage.

IV

The fourth illustration of the necessity for independence derives from our close cousins, university press publishers. Even great houses are touched by commercialism and tinged by fear. Our closest head-to-head competition in this area came from Cambridge University Press and Oxford University Press. I will hold for the penultimate "case study" our encounter with a Cambridge book, and focus here on David Stoesz's co-authored work on the National Association for Social Welfare and his critique of the social welfare network in the United States. The recording is familiar enough: the authors wrote a book that took dead aim at a profession of social welfare, tightly wrapped by an organization and virtually impervious to critical, much less self-critical, examination.

David Stoesz's book, *A Dream Deferred: How Social Work Education Lost Its Way and What Can Be Done About It*, was accepted, set in type, and circulated in proof by Oxford University Press early in 2009, only to be quietly and quickly cancelled prior to publication. It is clear that despite academic peer review, the directors of the National Association of Social Work were instrumental in seeing to it that the book would not be published. The implicit threat—there seem always to be such implicit threats, even if they are often just paper tigers without real teeth—was that books and journals published by Oxford would be banned, or at least treated severely, by the social welfare community upon such publication. This was a serious book, written by a person who had given a lifetime of dedicated service to the area of social work and who, indeed, had published an essay on this subject in *Society* as early as 1986.

Stoesz's book is well written; it is hardly an incendiary indictment in its frank criticism of the organizational cadre that determines policy in the field, as well as membership in its ranks; and the policies suggested for its intended victims are moderate. Here we have a case of a publisher and a professional agency teaming up to ensure that no criticism is publicly made that might threaten funding by government or ideology by recipients. I do not know, of course, how often such problems arise; I do know that they are hardly unique in the annals of academic publishing.

Indeed, *Alms for Jihad*, a book published by Cambridge and written by J. Millard Burr and Robert Collins, was offered to Transaction with the proviso that our lawyers give their approval.

One author notes that he "has found that while publishers may be interested in the book their lawyers are far less so." Transaction would have been prepared to move ahead with this title, but the proposed new lengthy introduction seemed far more concerned with a legal defense of the authors' position and their arguments with Cambridge than with expanding on the theme of the book itself: the Saudi banking houses and their connection with the jihadists.

The virtual collapse of university presses in the face of religious threat is not a monopoly of Cambridge. More recent is the 2009 decision of Yale University Press, one of the premier quality houses, not to include twelve Danish drawings in a book titled *The Cartoons That Shook the World*. Since the very purpose of the book is to highlight this specific set of illustrations, which appeared in 2005, the decision to suppress them—and also to obey anonymous recommendations that Yale refrain from publishing any other illustrations of the prophet Mohammed—is indicative of the state of affairs in even the best and strongest of American university presses. The author, Jytte Klausen, a Danish-born professor of politics at Brandeis University, while disturbed by the decision of the press, reluctantly accepted the ruling. Reza Aslan, an expert who had been asked to blurb the book for Yale, opted to withdraw his support after the press's director and board dropped the images. "The book is a definitive account of the entire controversy," he said; "but not to include the actual cartoons is to me, frankly, idiotic…. This is an academic book for an academic audience by an academic press. There is no chance of this book having a global audience, let alone causing a global outcry. It's not just academic cowardice, it is just silly and unnecessary." Be that as it may, Yale's decision to yield on this matter, as well as to place what the author has called "a gag order" on her sentiments and her book, indicates an ominous tendency in this precious scholarly world. John Donatich, the press director, said that "the decision was difficult, but the recommendation to withdraw the images, including the historical ones of Muhammad, was 'overwhelming and unanimous.'"

V

This coming up against the world of university presses poses a peculiar and special problem—one in which an author may be less the victim than the victimizer—not an easy role, or one to which professional readers are especially welcoming. And that brings us to the fifth case, the publication of a book accepted (but never issued) by Cambridge: Philip Rushton's *Race, Evolution and Behavior*. Professor Rushton, a strong-willed psychologist with a long pedigree of works on the same subject, submitted his book to Cambridge. It was accepted for publication, reviewed by no fewer than five experts in the field—each with grudging respect for the accuracy and scientific veracity of the text. The grudging part is clear enough: Rushton placed special emphasis on racial bipolarities in education results, disparities that continue to haunt the field of education—as with the educational success stories at Chicago public schools that perpetuated the myth that levels of education are greatly improving, real evidence to the contrary notwithstanding.

Transaction published this book in 1994, with the proviso of enlisting several additional expert opinions on the text, one of which prophetically said, "Publish this book and then run for the hills." We did just that—although we ran too slowly, it seems. The published book received extremely favorable front-page reviews in the *New York Times* and in the pages of *Scientific American*. The book itself did relatively well, despite problems with the text. Indeed, I was in the uncomfortable position of writing three different critiques of Rushton and his book, despite our having published the work. These appeared in *Society*, of which I was editor, and require little embellishment at this point or in this context.

But that was not the end of the story. Rather, it was Professor Rushton who managed to convert a victory into a defeat—to the satisfaction of those who wished nothing more than that such a book had never been published.

Rushton's book was successful, in good measure, because both he as author and we as publisher built a high wall between psychological truths and policy fantasies. The author decided to break that wall, and in the final year of the last century issued a postscript in pamphlet form—making plain his racially biased belief that the gulf in the capacity for learning between the races was virtually fixed by inheritance. He managed, in his mind at least, to soften the blow by claiming that "Asians" were highest on the biological learning curve, blacks the lowest, and whites somewhere in a vast purgatorial middle. This pamphlet, which also included an appeal for funds for a Darwin Foundation that Rushton had set in motion, was improperly characterized as an "abridged" version of the book, and widely disseminated as such. Needless to say, the outrages that followed were loud and ceaseless. They reached the portals of the Rutgers University officials, no less than our own.

We were faced with a conundrum: an author who insisted that the original book and his revisionist pamphlet were one and that any attempt to suppress its distribution was an attack on the First Amendment rights of a righteous Canadian. This position was repeatedly put forth by Rushton at places like the National Press Association in Washington, DC, and points south. At the other end were university officials wondering about the limits of free speech against a massive display of prejudice, given that the needs of the people of New Jersey included the advancement of African Americans in terms of student and faculty numbers, as well as through programs aimed precisely at maximizing equitable outcomes. This was not a simple matter, and raised, as happens often in the life of publishing, a sense of the unending struggle of what Goethe called the struggle between the good and the good.

It was my feeling that the work could no longer continue to be published by Transaction—that the wall that separates science and ideology, scientific analysis and racial bigotry, had been severely breached. We then negotiated an agreement that returned the rights to the book to the author and suspended any further sales or publicity on its behalf. I confess that this was probably the most trying moment I have had with my own belief in the right of a free nation to a free press. There were other such "tests," of a less momentous nature, for our house. I wish that the author had not broken the firm seal that exists between science and myth-making. I wish, too, that our university officials had been less taken with making clear that Transaction was not part of the university system, and hence not subject to lawsuits, and more appreciative of just how delicate this matter was for all concerned.

This decision was made not to change or alter the text but, rather, to curtail a dubious use of a serious scientific treatise for strictly ideological and promotional purposes. In retrospect, I believe, more than a decade later, that the decision to terminate this text in our list was correct—indeed, that it upheld the Weberian separations of facts and values. For while a publisher has a right to publish, it also has an obligation to protect its constituency against slander and defamation, based on the constitutional rights of all citizens. This episode taught me a personal lesson: that conceit and arrogance can exist on all levels; among those puffed up over unfurling the banner of self-righteousness, no less than among those who may be overly conscientious in protecting citizens from access to differences of opinion, attitudes, and information.

VI

There is irony in these test-case illustrations of a publishing world laden with danger and risk as well as reward and glory. One such item is the case of Laud Humphries and his brilliant dissertation *Tearoom Trade*—a major study of the 1960s on homosexual behavior in public places, with a particular emphasis on the social backgrounds and characteristics of its participants. Again, the issues are both well documented and reasonably well known without a recitation of the circumstances. The dissertation was accepted at Washington University in 1968 only on condition that for two years it would not be published or its contents revealed to a wider public. Whatever the causes and reasons for such a bizarre decision, and the collusions of a professional department within the university with university officials, this deal was struck and its conditions observed by all parties.

At the time, I was serving as external advisor in the social sciences for Oxford University Press. I strongly urged that great publishing house to issue the book once the two-year window expired. The reviews of the manuscript were excellent, but my concerns that a major piece of research, however volatile, coming at a time of increasing noise and decreased learning on the subject of homosexual behavior in American life, merited publication, despite some troubling aspects of the methodology in research procedure, were rejected by the Press. For reasons that were not ever made public or fully expressed, the decision was made not to publish Humphries' text. I had no choice but to accept this decision, but it also led to my accelerating the development of a book division at Transaction, and hence to a quiet resignation from the Oxford position in the name of avoiding a conflict of publishing interests.

The upshot of this was that *Tearoom Trade: Impersonal Sex in Public Places* was published by Aldine-Atherton Publishers in Chicago in 1970, to the lasting credit of Alex Morin, director of that house. The book did very well, the author achieved the notoriety and fame that he properly desired and merited, and the interests of the public were served—albeit not easily or readily. Controversy continued, and in the pages of *Society* a bitter controversy between the journalist Nicholas von Hoffman and me took place. The issues revolved around the ethics of the author rather than the cowardice of the publishing and professional environment, which at the time was fixated on establishing "professional ethics" as legal guidelines. Such concerns are now part of the environment, part of the struggle for a free press in the domain of higher learning.

The ironic part of the story came later, in 2004, long after Laud Humphries' book had become a staple in the literature of social research. Transaction purchased Aldine, which several decades since had passed from an independent publisher to one owned by De Gruyter Publishing in Germany; as part of the purchase, *Tearoom Trade* became the property of Transaction. The book continues in print, and it retains a place in the field of sociological classics. Laud, who was my student and on whose dissertation committee I served as second in command to Lee Rainwater, a colleague at Washington University in the 1960s, had a new home for his work. And while this title had a happily-ever-after ending, it is one mitigated by the author's tragic death in 1987 as a result of AIDS. But *Tearoom Trade* remains a beacon in the struggle for a free and independent social science publishing community.

With this bittersweet personal coda to the larger issue of what it means to struggle for a free press, I bring to a close this story of publishing as a vocation.

Conclusion

This set of "case studies" reveals a preponderance of instances in which our house has come down on the side of free expression and untrammeled writing on important social science themes. But it also serves to remind us of the great merit of Max Weber's distinctions between the truths of empirical research and the goals of political reform. Over time, and in the fullness of discussion, the essential truths of this Weberian theory of the duality of our commitment to the search for truth and the desire for a better world for which ideologies prevail will be tested. It is far better that difficult issues be decided in the court of public opinion than that they be circumvented by intellectual fiat. Ultimately, this is what sets publishing houses like ours apart from authorities—whether of governments, professional agencies, or publishers weighed down by tradition and cautious academic advisors more intent on censorship than on scholarship. In such publishing, there are few formulas to advance. But there is a tradition of science in research and decency in human relations that can guide us. Onward.

References

Jitu Brown, Eric Gutstein, and Pauline Lipman, "Arne Duncan and the Chicago Success Story: Myth or Reality," *Rethinking Schools* 23, 3 (Spring 2009): http://www.rethinkingschools.org/archive/23_03/arne233.shtml.

Patricia Cohen, "Yale Press Bans Images of Muhammad in New Book," *New York Times* (August 13, 2009): C1.

David Stoesz, Correspondence with Transaction Publishers, June 25, June 30, and July 17, 2009.

Max Weber, "Politics as a Vocation" and "Science as a Vocation," *From Max Weber: Essays in Sociology*, eds. H.H. Gerth and C. Mills, (New York: Oxford University Press 1946), 77-128 and 129-58.

Index